Landmarking and Segmentation of 3D CT Images

Synthesis Lectures on Biomedical Engineering

Editor
John D. Enderle, *University of Connecticut*

BioNanotechnology
Elisabeth S. Papazoglou and Aravind Parthasarathy
2007

Bioinstrumentation
John D. Enderle
2006

Fundamentals of Respiratory Sounds and Analysis
Zahra Moussavi
2006

Advanced Probability Theory for Biomedical Engineers
John D. Enderle, David C. Farden, and Daniel J. Krause
2006

Intermediate Probability Theory for Biomedical Engineers
John D. Enderle, David C. Farden, and Daniel J. Krause
2006

Basic Probability Theory for Biomedical Engineers
John D. Enderle, David C. Farden, and Daniel J. Krause
2006

Sensory Organ Replacement and Repair
Gerald E. Miller
2006

Artificial Organs
Gerald E. Miller
2006

Signal Processing of Random Physiological Signals
Charles S. Lessard
2006

Image and Signal Processing for Networked E-Health Applications
Ilias G. Maglogiannis, Kostas Karpouzis, and Manolis Wallace
2006

Landmarking and Segmentation of 3D CT Images

Shantanu Banik, Rangaraj M. Rangayyan, and Graham S. Boag

ISBN: 978-3-031-00507-7   paperback
ISBN: 978-3-031-01635-6   ebook

DOI 10.1007/978-3-031-01635-6

A Publication in the Springer series
SYNTHESIS LECTURES ON BIOMEDICAL ENGINEERING

Lecture #30
Series Editor: John D. Enderle, University of Connecticut

Series ISSN
Synthesis Lectures on Biomedical Engineering
Print 1932-0328   Electronic 1932-0336

Landmarking and Segmentation of 3D CT Images

Shantanu Banik
University of Calgary

Rangaraj M. Rangayyan
University of Calgary

Graham S. Boag
Alberta Children's Hospital

SYNTHESIS LECTURES ON BIOMEDICAL ENGINEERING #30

ABSTRACT

Segmentation and landmarking of computed tomographic (CT) images of pediatric patients are important and useful in computer-aided diagnosis (CAD), treatment planning, and objective analysis of normal as well as pathological regions. Identification and segmentation of organs and tissues in the presence of tumors are difficult. Automatic segmentation of the primary tumor mass in neuroblastoma could facilitate reproducible and objective analysis of the tumor's tissue composition, shape, and size. However, due to the heterogeneous tissue composition of the neuroblastic tumor, ranging from low-attenuation necrosis to high-attenuation calcification, segmentation of the tumor mass is a challenging problem. In this context, methods are described in this book for identification and segmentation of several abdominal and thoracic landmarks to assist in the segmentation of neuroblastic tumors in pediatric CT images.

Methods to identify and segment automatically the peripheral artifacts and tissues, the rib structure, the vertebral column, the spinal canal, the diaphragm, and the pelvic surface are described. Techniques are also presented to evaluate quantitatively the results of segmentation of the vertebral column, the spinal canal, the diaphragm, and the pelvic girdle by comparing with the results of independent manual segmentation performed by a radiologist.

The use of the landmarks and removal of several tissues and organs are shown to assist in limiting the scope of the tumor segmentation process to the abdomen, to lead to the reduction of the false-positive error, and to improve the result of segmentation of neuroblastic tumors.

KEYWORDS

medical image analysis, computed tomography (CT), computer-aided diagnosis (CAD), three-dimensional (3D) image processing, landmarking, image segmentation, atlas-based segmentation, tumor segmentation, fuzzy region growing, morphological image processing, opening-by-reconstruction, active contours, vertebral column, rib structure, spinal canal, diaphragm, pelvic girdle, neuroblastoma

Shantanu Banik dedicates this book to
his parents Pradip Kumar Banik and Anuradha Banik,
and to his wife Shormistha Prajna.

Rangaraj M. Rangayyan dedicates this book to
Dr. Richard Gordon, University of Manitoba, Winnipeg,
Manitoba, Canada,
for introducing him to CT.

Contents

Preface

This book describes several image processing techniques, both classical and fuzzy, and their applications in segmentation and landmarking of medical images. Segmentation of the organs in computed tomographic (CT) images could aid radiologists by providing reproducible and objective assessment. However, due to the heterogeneous nature of the tissues, lack of clear boundaries, similarities among the adjacent organs, noise, partial-volume effect, and limitations of the imaging systems, accurate segmentation of the organs is a difficult problem. In this context, knowledge-based, atlas-based, and landmark-based image processing techniques to perform identification, localization, and segmentation of several organs and tissues in pediatric CT images are presented in this book. In addition, the application of landmarking in segmentation of abdominal neuroblastic tumors of pediatric patients is also described.

The success of segmentation of the neuroblastic tumor depends largely on the location of the tumor in the abdomen and the tissue composition of the tumor. Due to heterogeneity of the tumor, the result of segmentation includes leakage into other abdominal organs. Despite the fact that many approaches to segmentation of abdominal organs and tumors have shown promising results, accurate segmentation of abdominal organs is still a challenge, and there exists a need for further techniques and improvement from the present situation. Specifically, the liver, the kidneys, and several other organs are difficult to segment satisfactorily even after considering the features of neighboring areas and using prior knowledge in the form of an atlas.

This book is intended for engineering students and professionals working on medical images and on the development of systems for computer-aided diagnosis (CAD). The methods and technical details presented in this book are at a fairly high level of sophistication, and should find use in medical image processing for practical applications.

Acknowledgments

This book is the result of the research conducted by and contributions from many individuals and organizations. We gratefully acknowledge Randy H. Vu, Hanford Deglint, and Dr. Fábio J. Ayres for their enthusiastic efforts, hard work, enormous contribution, and support in the preliminary stages of this research. We thank the Natural Sciences and Engineering Research Council (NSERC) of Canada, the Alberta Heritage Foundation for Medical Research (AHFMR), Kids Cancer Care Foundation (KCCF) of Alberta, the University of Calgary, and the Alberta Children's Hospital for the resources provided.

We thank the Institute of Electrical and Electronics Engineers (IEEE), the IEEE Engineering in Medicine and Biology Society (EMBS), the Canadian Conference on Electrical and Computer Engineering (CCECE), the Society of Photo-Optical Instrumentation Engineers (SPIE), the Society of Imaging Informatics in Medicine (SIIM), International Journal of Computer Assisted Radiology and Surgery (IJCARS), and Springer Science+Business media for their support and permission to use published material.

Some of the materials and illustrations have been reproduced, with permission from the associated organizations, and are listed below:

1. R M Rangayyan, S Banik, and G S Boag. "Landmarking and segmentation of computed tomographic images of pediatric patients with neuroblastoma." *International Journal of Computer Assisted Radiology and Surgery*, 2009. In press. © Springer.

2. S Banik, R M Rangayyan, and G S Boag. "Automatic segmentation of the ribs, the vertebral column, and the spinal canal in pediatric computed tomographic images." *Journal of Digital Imaging*, 2009. In press. © SIIM.

3. S Banik, R M Rangayyan, G S Boag, and R H Vu. "Segmentation of the pelvic girdle in pediatric computed tomographic images." Submitted to the *Journal of Electronic Imaging*, June 2008. © SPIE.

4. S Banik, R M Rangayyan, and G S Boag. "Landmarking of computed tomographic images to assist in segmentation of abdominal tumors caused by neuroblastoma." In *Proceedings of the 30th Annual International Conference of the IEEE Engineering in Medicine and Biology Society*, pages 3126–3129, Vancouver, BC, Canada, August 2008. © IEEE.

5. R M Rangayyan, S Banik, and G S Boag. "Automatic segmentation of the ribs and the vertebral column in computed tomographic images of pediatric patients." In *Proceedings of CARS 22nd International Congress and Exhibition: Computer Assisted Radiology and Surgery*, volume 3(1), pages S42–S44, Barcelona, Spain, June 2008. © Springer.

6. S Banik, R M Rangayyan, and G S Boag. "Delineation of the pelvic girdle in computed tomographic images." In *Proceedings of the 21st IEEE Canadian Conference on Electrical and Computer Engineering*, pages 179–182, Niagara Falls, Ontario, Canada, May 2008. © IEEE.

We thank our families, relatives, and friends for their good wishes, inspiration, patience, appreciation, and loving care.

Shantanu Banik,
Rangaraj M. Rangayyan,
Graham S. Boag

Calgary, Alberta, Canada
March, 2009

List of Symbols and Abbreviations

$\mathcal{A}$	fuzzy set
B	morphological structuring element
$\check{B}$	reflection of B
$\mathbf{C}$	closed curve
$C_{m_{\mathcal{A}}}(p, q)$	degree of connectedness between p and q
CAD	computer-aided diagnosis
CT	computed tomography
cf	compactness factor
D	image domain
E	energy functional
ED	Euclidean distance
$f(x, y)$	digital image
$\mathbf{F}$	Force
$g(x, y)$	binary image
GVF	gradient vector flow
G_x	gradient in the x-axis direction
G_y	gradient in the y-axis direction
h	morphological translation
h, H	Hausdorff distance
HU	Hounsfield unit
I	image mask
J	image marker
l	path or number of available gray levels
$L_{p,q}$	set of all paths from p to q
$L = 2^k$	number of gray levels
LoG	Laplacian of Gaussian
$m_{\mathcal{A}}$	fuzzy-membership function
MDCP	mean distance to the closest point
MRI	magnetic resonance imaging
$\mathcal{P}$	logical predicate
$\mathbb{R}^n$	n-dimensional real Euclidean space
RMS	root mean-squared

ROI	region of interest
SD	standard deviation
SNR	signal to noise ratio
T	threshold
$\mathcal{X}$	reference set
$\mathbb{Z}^n$	n-dimensional discrete space
p, q	pixels or voxels
(x, y)	Cartesian coordinate in xy plane
(x, y, z)	Cartesian coordinate in three-dimensional space
$\emptyset$	null or empty set
$2D$	two dimensions
$3D$	three dimensions
α	tension constant for deformable contour model
α	morphological translation
β	rigidity constant for deformable contour model
β	morphological translation
γ	viscosity constant for deformable contour model
$\delta^{(1)}$	elementary geodesic dilation
$\partial f / \partial x$	partial derivative in the x direction
$\partial f / \partial y$	partial derivative in the y direction
ε	error rate
ζ	specific mass of contour
η	strength of path
η_κ	strength of fuzzy connectedness
κ	fuzzy spel affinity
λ	smoothness regularization parameter for deformable contour model
μ	mean or average
μ	linear attenuation coefficient
∇	gradient operator
ρ_I	morphological reconstruction of I
σ	standard deviation
Ψ	set operator
$\forall$	for all
$\in$	belongs to or is in (the set)
$\sim$	complement
$\{\}$	set
$\subseteq$	subset
$\cap$	intersection
$\cup$	union

$\wedge$	point-wise infimum
$\vee$	point-wise supremum
$\mapsto$	maps into
$\Rightarrow$	implies
()	open interval, not including the limits
[ ]	closed interval, including the limits
$\oplus$	morphological dilation
$\ominus$	morphological erosion
•	morphological or structural closing
○	morphological or structural opening

CHAPTER 1

Introduction to Medical Image Analysis

1.1 MEDICAL DIAGNOSTIC IMAGING

The field of medical imaging and image analysis has evolved as a result of the collective contributions from many areas of medicine, engineering, mathematics, and basic sciences. The overall objective of medical imaging is to acquire useful information about the organs of the body as well as the ongoing physiological and pathological processes by using external or internal sources of energy [1, 2]. Imaging methods available currently for radiological imaging may use internal, external, or a combination of energy sources [2].

Due to varying physiological or pathological conditions, it is sometimes required to look inside the human body. The direct way to do this is to cut it open through surgery; the second way is to use such technologies or imaging techniques that permit the physician or radiologist to look inside the human body without cutting through the body or performing any surgery. Diagnostic imaging refers to technologies that allow physicians to examine the inner parts of the human body for clues about a medical condition. The choice of imaging modalities depends on the symptoms and the part of the body being examined. Cases may arise when several different types of imaging techniques may need to be incorporated to learn about the true condition of a particular organ of interest, or to localize pathological regions.

Several diagnostic imaging processes are painless, easy to apply, and noninvasive; however, they may require the patient to be held inside a machine, or demand the patient to maintain a certain condition, such as holding breath, performing an exercise, or remaining still in a certain position, for a period of time. Certain methods involve ionizing radiation and are generally considered to be safe because of controls on dosage. In some diagnostic imaging processes, cameras or other devices may be inserted into the body; they are considered to be invasive processes and may require sedation or anesthesia.

Many imaging modalities are in use today to acquire anatomical, physiological, metabolic, and functional information from the human body. Using computers, multi-dimensional digital images of anatomical structures can be processed and manipulated to visualize the diagnostic features that are difficult or impossible to see with planar imaging methods. In many critical radiological applications, multi-dimensional visualization and quantitative analysis of physiological structures provide clinical information that is extremely valuable in diagnosis and treatment. Applications such as early detection of breast cancer demonstrate the functionality, scope, and effectiveness of computer-aided diagnosis (CAD) [3, 4].

1.1.1   IMAGING MODALITIES

Several imaging modalities have been developed and used extensively to study and to identify different types of diseases. The most common imaging techniques that are in use at present are [2, 3, 5]:

- radiography, including mammography, X-ray angiography, transmission X-ray computed tomography (CT);

- ultrasonography (US);

- nuclear medicine imaging, including meta-iodobenzyl guanidine (MIBG) scintigraphy, positron emission tomography (PET), and single-photon emission computed tomography (SPECT); and

- magnetic resonance imaging (MRI).

In some commonly used imaging methods, such as X-ray imaging and CT, radiation is used as an external energy source primarily for anatomical imaging. Such anatomical imaging modalities are based on the attenuation coefficient of X rays passing through the body. CT imaging is a popular and useful tool in clinical applications because of the ability to provide improved contrast resolution at low levels of structured noise.

MRI is a technique that is considered to be noninvasive, painless, and well tolerated. MRI does not involve the use of ionizing radiation, and for many medical situations, it has become the modality of choice [6, 7]. In the case of detection of tumors, MRI may be more accurate than CT; however, acquisition times for CT are substantially less than those of MR scans [8]. CT scans are more cost effective and are able to detect calcification, which is highly desired in the detection of tumors. A major disadvantage of MRI is the requirement to sedate the pediatric patient [9]. In some cases, physicians or radiologists prefer to perform the evaluation and analysis of patients with both CT and MRI [10].

Diagnostic imaging is an integral aspect of the treatment and management of patients with tumors, cancer, and other pathological conditions, and permits noninvasive diagnosis of the disease. Diagnostic imaging comprises a variety of imaging modalities ranging from fluoroscopy and ultrasound to X-ray CT and MRI, which may be used in conjunction with one another to provide complementary information to aid in diagnosis. For example, in the treatment of neuroblastoma [11, 12, 13], a patient may be initially screened using ultrasound, and subsequently imaged using CT to guide the staging and treatment of the disease [14].

1.1.2   TISSUE CHARACTERIZATION IN CT IMAGES

In CT images, X rays are transmitted through the body to obtain projection measurements. As a given X-ray beam traverses through the body, it is attenuated exponentially according to the Lambert-Beer law, given by

$$I_t = I_o \exp(-\mu l), \tag{1.1}$$

where I_t denotes the transmitted intensity of the X ray, I_o represents the incident intensity [15], μ is the linear attenuation coefficient (assumed to be constant for the path considered), and l is the length of the path of the beam. According to Equation 1.1, as the length of the path or the attenuation coefficient of the object increases, the X ray will suffer increased attenuation.

Multiple projections around the body are acquired in CT, and by using a reconstruction algorithm, a three-dimensional (3D) representation consisting of equally spaced transversal slices of the body is obtained [5]. The physical characteristic represented in the reconstructed 3D dataset is μ; the value of μ varies according to a material's density and atomic composition, as well as the X-ray energy used [16]. To represent μ in a more convenient manner and to make it effectively independent of the X-ray energy used, it is converted to a CT unit that is normalized with respect to the attenuation coefficient of the reference material, water, defined as

$$\text{CT number} = k\,\frac{\mu - \mu_w}{\mu_w},\tag{1.2}$$

where μ_w is the linear attenuation coefficient of water and k is an arbitrary constant. The value of k is usually set to 1,000 to obtain CT numbers in terms of *Hounsfield Units* (HU) [15]. CT values have been tabulated for several tissues and organs in the body [17]; some examples are given in Table 1.1.

In order to improve the definition of organs or pathological regions in CT, intravascular contrast agents are usually employed in the imaging study. The contrast agent is injected rapidly into the venous system, with scanning commencing shortly after completion of the injection. The contrast agent concentrates preferentially in tumoral or pathological tissues, thus maximizing the density difference between the abnormal regions and the surrounding tissues, and allowing for enhanced definition of the region. Several of the HU values in Table 1.1 were estimated using CT data with contrast.

In the detection and segmentation methods described in the following chapters, HU values of CT voxels are an integral and significant basis of the associated algorithms. Many of the imposed restrictions in the strategies presented are directly dependent on the expected HU value of the specific organ or tissue being segmented or analyzed.

1.2   COMPUTER-AIDED ANALYSIS OF MEDICAL IMAGES

Radiologists, physicians, cardiologists, neuroscientists, pathologists, and other medical professionals are highly skilled in the analysis of visual patterns in medical signals and images. Although manual analysis and estimation can provide accurate assessment of the organs of interest and any associated pathological conditions, the methods are generally tedious, time-consuming, and subject to error. Analysis by humans is usually subjective and qualitative. The accuracy achievable is limited mainly by the ability to define a precise boundary of the region of interest (ROI) in medical images. In order to perform objective or quantitative, consistent, and fast analysis, and to help the medical professionals in the decision-making process, image processing techniques are widely used for computer-aided analysis of medical images. CAD is a concept established by taking into account the roles of physicians

Table 1.1: Mean and standard deviation (SD) of the CT values of several organs and tissues in Hounsfield Units (HU).

Tissue	CT value (HU) mean	SD
Air[†]	-1006	2
Fat*	-90	18
Bile*	+16	8
Kidney*	+32	10
Pancreas*	+40	14
Blood (aorta)*	+42	18
Muscle*	+44	14
Necrosis[†]	+45	15
Spleen*	+46	12
Liver*	+60	14
Viable tumor[†]	+91	25
Marrow*	+142	48
Calcification[†]	+345	155
Bone[†]	+1005	103

Note: This table has been reproduced with permission from F J Ayres, M K Zuffo, R M Rangayyan, G S Boag, V Odone Filho, and M Valente. "Estimation of the tissue composition of the tumor mass in neuroblastoma using segmented CT images." *Medical and Biological Engineering and Computing*, 42:366–377, 2004. © Springer.
*Based on Mategrano et al. [18].
[†]Estimated from CT examinations with contrast. The contrast medium is expected to increase the CT values of vascularized tissues by 30 – 40 HU.

and computers; the physicians use the computer output as a "second opinion" to make the final decision [3, 4].

Accurate identification and segmentation of abdominal organs in CT images are important in CAD systems to improve diagnostic accuracy, reduce computational cost, and localize pathological regions [19, 20, 21, 22, 23, 24]. However, due to the wide range of variability of the position, size, shape, texture, and gray-level values of organs and tissues; partial-volume effects; the heterogeneous nature of different organs; similarities of neighboring organs; and the effects of contrast media, the automatic identification and segmentation of any abdominal organ is a highly challenging problem. Prior knowledge, anatomical atlases, anatomical landmarks, and relative coordinate systems have been used for improved localization, identification, and registration of various organs [19, 20, 25, 26].

1.2.1   KNOWLEDGE-BASED SEGMENTATION

Computerized processing and analysis of medical images of different modalities provide powerful tools to assist physicians and radiologists. The methods to improve the diagnostic information in medical images can be further enhanced by designing computerized methods that are able to make the best use of available prior information. Incorporating relevant knowledge, such as the physics of imaging, instrumentation, human physiology and anatomy, and the geometrical position of the patient often provides improvement in the analysis of medical images [2].

Knowledge-based segmentation algorithms use explicit anatomical knowledge, such as the expected size, shape, and relative positions of objects. This knowledge is used to guide low-level image processing routines to achieve better discrimination between objects of similar characteristics [27]. In computer-based systems for the analysis of medical images, structural knowledge, procedural knowledge, and/or dynamic knowledge regarding the possible normal and abnormal processes may be incorporated in the segmentation or recognition process [28]. Such knowledge can be incorporated by defining rules, using semantic networks, developing anatomical atlases, modeling anatomical variability, or using relative position with respect to landmarks [27]. With new advances in image processing, adaptive learning, modeling of organs of interest, and knowledge-based image analysis techniques can help to improve the performance of CAD systems [2, 27].

Knowledge-based segmentation can assist in delineating anatomical objects of similar characteristics more reliably, and thus reduce operator intervention. The associated processes may be automatic or semi-automatic, and have been used for the segmentation of abdominal and thoracic organs [20, 25, 29].

1.2.2   ATLAS-BASED SEGMENTATION

An atlas can be an anatomical map or image of an individual, or an average map or image of multiple individuals in a specific group. An atlas contains knowledge of the spatial and density distributions of anatomical structures. Segmentation accuracy can be greatly improved by using multiple atlases [30, 31, 32]. Digital atlases have been used for registration, localization, and segmentation of organs and structures in the body, as well as for assistance in the planning of surgery [30, 31, 33, 34]. In addition, the construction of probabilistic atlases [19] can help in the extraction of single or multiple abdominal organs.

A number of atlas-based segmentation algorithms have been proposed [19, 21, 22, 23, 24, 32, 35, 36, 37]; such procedures require *prior* knowledge of the atlas, along with knowledge of the expected size, shape, texture, and position of the organ of interest relative to one or more landmarks. Atlas-based segmentation approaches may use a variety of techniques, such as neural networks, model fitting, level sets, deformable models, rule-based recognition, and data-directed methods.

There are difficulties in representing anatomical variations using an atlas; abnormal anatomical structures, such as tumors, cannot readily be incorporated into an atlas, because their size, shape, and location are highly variable. It is not possible to label a set of pixels or voxels in a generic atlas as a tumor because the location of a tumor would be different in another patient: this prevents transformation of the atlas to fit a specific patient's data [27].

1.2.3   LANDMARKING OF MEDICAL IMAGES

Landmarks are used as references to describe the spatial relationship between different regions, organs, or structures in medical images. In the case of abdominal and thoracic CT scans, landmarks are selected in such a manner that they are easy to detect, have stable locations, and have characteristics that do not alter to a great extent in the presence of abnormalities. Usually, the spine, the spinal canal, the rib structure, the diaphragm, and the pelvis are used as landmarks [22, 28, 38, 39, 40, 41]. In cases where there exist pathological or abnormal regions, and segmentation relying solely upon atlas-based information is likely to fail, landmark-based image segmentation may be incorporated in conjunction with other knowledge-based procedures. The use of reliable landmarks can accommodate variations in anatomical structures, and thus help to make the process flexible.

Various algorithms have been developed to identify abdominal landmarks [26, 28, 38, 39, 40, 41, 42, 43, 44], to aid atlas-to-subject registration, or to perform segmentation of an organ of interest. The locations of the spine and the thoracic cage have been used for the purpose of landmarking in the segmentation of various abdominal and thoracic organs [23, 39]. Shimizu et al. [22, 36] proposed the use of five landmarks, including the upper-end voxels of the diaphragm and the kidneys, and the lower-end voxels of the liver and spleen, to perform the segmentation of multiple abdominal organs in CT images.

In image processing techniques for practical applications, knowledge-based approaches often use atlas-based information and several landmarks to perform computer-aided segmentation, diagnosis, registration, and other related analysis.

1.3   OBJECTIVES AND ORGANIZATION OF THE BOOK

As discussed in the preceding sections, several works have been directed toward the development of anatomical and probabilistic atlases for adults [19, 21, 23, 24, 32, 35, 36, 37, 45] and the segmentation of tumors [46, 47]. Kim and Park [46] used information related to the spine and the kidneys to perform segmentation of renal tumors, and achieved a sensitivity of 85% with no false positives. Qatarneh et al. [31] presented a semi-automated technique for the segmentation of multiple organs to assist in accurate tumor localization and radiation therapy. Kaus et al. [48] achieved accuracy in the range of 99.6% ± 0.29% in automatic segmentation of brain tumors using brain atlases and several segmented landmarks in magnetic resonance (MR) images. Soler et al. [49] reported success in the segmentation of hepatic tumor lesions in CT images with automatic prior delineation of the skin, bones, the lungs, the kidneys, and the spleen by combining the use of thresholding, mathematical morphology, and distance maps.

However, very few works have been reported for the pediatric age group to assist in the segmentation of abdominal organs and pathological regions, such as tumors. The major problems in building atlases for children are: they vary to a larger extent than adults in terms of the shape, size, texture, and position of organs; some of the organs may not be fully developed, depending on the age and ethnicity; and they demonstrate characteristics that are different from those of adults in CT images. In addition, the presence of pathological regions or tumors could displace and deform some of the abdominal organs, which could create more ambiguity with respect to other anatomical structures. As a result, few reliable landmarks may exist in pediatric CT images to aid registration or segmentation of organs or tumors, or to facilitate the use of standard atlases.

The accurate localization and diagnosis of abdominal tumors, such as neuroblastoma [50, 51, 52], is a challenging problem. Segmentation and analysis of the primary tumor mass in neuroblastoma could aid radiologists by providing reproducible and objective quantification of the tumor's tissue composition and size. However, due to the heterogeneous nature of the tissue components of the neuroblastic tumor [13, 53], ranging from low-attenuation necrosis to high-attenuation calcification, some of which possess strong similarities with adjacent nontumoral tissues, proper segmentation of the tumor is a difficult problem.

In this context, Rangayyan et al. [42] proposed methods for automatic identification of the spinal canal to facilitate the segmentation of neuroblastic tumors [11, 12]. Vu et al. [12] and Rangayyan et al. [54] proposed procedures for segmentation of the peripheral muscle and for identification of the diaphragm, which could be used as effective landmarks for the identification of several abdominal organs. Although the results of segmentation of tumors showed promise, further work is required to minimize leakage of the segmented region into the lower abdomen and to facilitate the delineation of other abdominal organs. In this context, the use of the pelvic girdle [55, 56, 57, 58], the rib structure, and the vertebral column [59, 60, 61] are also studied in the landmarking and segmentation of CT images.

With the issues described above as the motivating factors, the objectives of this book are to study 3D image processing methods for automatic identification and segmentation of several landmarks in pediatric CT images [55, 56, 57, 59, 60, 61], and to explore the use of the landmarks in the segmentation of abdominal tumors due to neuroblastoma [56, 59].

This book is organized into eight chapters. Chapter 2 provides a brief review on image processing, image segmentation, and image analysis as related to the methods described in the book. Commonly used procedures for image segmentation, such as thresholding, region growing, and edge-based techniques are described, and a brief outline of morphological image processing is provided. In addition, advanced image processing procedures, such as fuzzy connectivity and deformable models are briefly discussed.

The experimental design, the datasets, and the methods used for the evaluation of the results of segmentation are presented in Chapter 3.

In Chapter 4, the preprocessing steps, such as the procedures for delineation of the peripheral artifacts, the peripheral fat, and the peripheral muscle, are presented. Then, methods for automatic identification and segmentation of the rib structure, the vertebral column, and the spinal canal are described. Quantitative and qualitative evaluation of the results of segmentation are provided.

The process of delineation of the diaphragm using information related to the segmented lungs is presented in Chapter 5. The chapter is concluded with qualitative and quantitative assessment of the results. Because the diaphragm is the separating boundary between the abdomen and the thorax, it can be used to separate the abdominal cavity from the thoracic cavity.

In Chapter 6, the methods used for segmentation of the pelvic girdle and the procedures to obtain the upper surface of the pelvic girdle are discussed. The pelvic girdle is an important landmark in CT images, and the upper surface of the pelvic girdle can be used to assist in the localization and segmentation of other abdominal or pelvic organs. The results of the process of delineation of the pelvic girdle are assessed qualitatively and quantitatively, and the associated measures are presented at the end of the chapter.

The use of the identified and segmented landmarks in the problem of segmentation of the neuroblastic tumor is discussed in Chapter 7. In this chapter, information regarding the neuroblastic tumor and the strategies to achieve tumor segmentation are discussed. As part of the procedure, several known structures within the body are delineated and removed from consideration; the remaining parts of the image are then examined for the tumor mass. In addition, a description outlining the procedures used to achieve tumor segmentation is given; several comparisons are made, mentioning the benefits of each step of the described procedures. The results of segmentation are compared with the tumor definition as specified by a radiologist.

Chapter 8 concludes the book with a summary, and provides suggestions for potential future work.

CHAPTER 2

Image Segmentation

2.1 DIGITAL IMAGE PROCESSING

An image may be defined as a continuous two-dimensional (2D) light-intensity function $f(x, y)$, where x and y are the spatial (plane) coordinates such that $(x, y) \in \mathbb{R}^2$, and the amplitude of f at any pair of coordinates (x, y) denotes the *intensity* or *gray level* of the image at that point with $f \in \mathbb{R}$. Here, $\mathbb{R}^n$ denotes the n-dimensional real Euclidean space. For a 3D image, the function may be defined as $f(x, y, z)$, where the coordinates are from $\mathbb{R}^3$ and $f \in \mathbb{R}$. Generally, higher values of light intensity correspond to brighter elements in the image. When x, y, z, and f are discrete variables, the image is called a *digital image*.

In order to perform computer-based operations on an image, $f(x, y)$ must first be digitized by sampling the spatial coordinates and quantizing the corresponding amplitude values. As a result of digitization, the image is defined on the digital image domain D, where $D \in \mathbb{Z}^n$, and $\mathbb{Z}^n$ is the n-dimensional discrete space [62]. In other words, a digital image is composed of a finite number of elements, each of which has a particular location and value; the picture elements of a 2D image are referred to as *pixels* or *pels*. In the case of 3D images, the volume elements are called *voxels*. The field of *digital image processing* refers to the processing of digital images by means of a digital computer [5, 62].

Sampling refers to the extraction of equally spaced samples from a continuous function to obtain discrete points; that is, sampling is the digitization of the coordinate values. For an image, sampling results in a grid-like approximation of $f(x, y)$, which may be treated as a matrix with dimensions $M \times N$. Consequently, sampling is the principal factor determining the *spatial resolution* or the smallest discernible spatial detail in an image.

The second step in image digitization is *quantization*, which is a mapping of the values in the image into the set of available gray levels, $l = \{0, 1, ..., L - 1\}$, that can be represented on a digital computer. Depending on the number of bits, k, used to represent the pixel value, there are L discrete gray levels in which to map the continuous image values, with $L = 2^k$ [62]. By increasing the number of available quantization levels, the digitized image will more closely approximate the original image [5], thereby improving the *gray-level resolution*, or the degree of discernible detail in light intensity.

All images contain some level of degradation due to noise and unwanted artifacts, which arise from several different sources, including the image acquisition system, the environment of the experiment, and the natural physiology of the body in the case of medical images [5]. In CT images, noise and artifacts may arise due to beam hardening, scattering of X rays in the body, nonlinear tissue attenuation, limited detector efficiency, and motion artifacts from the patient [63]. Understanding these different forms of noise and artifacts is important because they affect the effectiveness of subsequent analysis.

Digital images may be manipulated in a variety of ways for many applications. The different manipulations can be categorized as *image processing* and *image analysis* [5]. Image processing aims to enhance an image by improving image quality or by augmenting specific desirable features. Image analysis, on the other hand, relates to the interpretation of the information within an image. Both categories of operations are important for the accurate and effective analysis of the information contained in the image, but there is no general agreement among authors in the area regarding where image processing stops and image analysis or computer vision starts [62]. However, digital image processing encompasses a wide and variety of applications.

Image enhancement is among the simplest and most appealing areas of digital image processing. It is a subjective area of image processing: the idea behind image enhancement is to bring out the details in the given image that are obscure or subtle, or to boost or highlight certain features of interest.

Segmentation, a form of image processing, is the process of partitioning an image into regions representing the different objects in the image. Segmentation algorithms for gray-scale images are based generally on one of two basic properties: *discontinuity* and *similarity* [5, 62, 64]. Discontinuity refers to the abrupt changes between the objects and the background in an image. On the other hand, similarity describes the uniformity and homogeneity within a given object or region in an image; a connected set of pixels having more or less the same homogeneous characteristics forms a region [64]. The first approach involves the detection of points, lines, and edges in an image, whereas segmentation methods based on the second property include thresholding, region growing, and region splitting and merging.

Segmenting objects in an image is a difficult task. Generally, the boundaries surrounding the objects in a medical image are subtle, or the pixels corresponding to an object of interest lack a sufficient level of similarity to achieve accurate segmentation. Furthermore, the partial-volume effect, noise, and artifacts degrade the image and interfere with the definition of regions and objects in 3D CT images.

Morphological image processing [65] deals with tools to extract image components that are useful in the representation of shapes. *Pattern Recognition* [66] is the process that assigns a label to an object based on its features or descriptors.

The purpose of this chapter is to review several approaches to image segmentation, both classical and fuzzy, and to discuss their benefits and drawbacks. Methods of thresholding, region-based approaches, edge-based techniques, deformable contour models, fuzzy connectivity, and morphological image processing are described in the following sections.

2.2   HISTOGRAM

The *dynamic range* of an imaging system or a variable is its range or gamut of operation, usually limited to the portion of linear response. The dynamic range of an image is usually expressed as the difference between the maximum and the minimum values present in the image, and gives concise information about the global range of the image intensity levels. On the other hand, the *histogram* provides detailed information on the number of pixels or voxels with a certain gray level over the complete dynamic range of the image.

The histogram of a digital image $f(x, y)$ of size $M \times N$ with L gray levels is defined as [5]

$$P_f(l) = \sum_{x=0}^{M-1} \sum_{y=0}^{N-1} \delta[f(x, y) - l], l = 0, 1, 2, \ldots, L - 1, \tag{2.1}$$

where the discrete delta function $\delta(n)$ is defined as [67]

$$\delta(n) \equiv \begin{cases} 1, & \text{for } n = 0 \\ 0, & \text{otherwise.} \end{cases} \tag{2.2}$$

The sum of all the entries in the histogram will equal the total number of pixels or voxels in the image. It is a common practice to normalize the histogram by dividing each entry in the histogram by the total number of pixels or voxels. Assuming that the image has a large number of pixels (or voxels), the histogram will approximate the probability density function (PDF) of the gray levels in the image.

As an example, the histogram of the image shown in part (a) of Figure 2.1, is displayed in part (b) of the same figure. The histogram is displayed only for the range $[-800, 800]$ HU out of the range $[-3024, 1071]$ HU present in the image. The largest peak around +100 HU corresponds mostly to the liver and the spleen in the image.

2.3   THRESHOLDING

A popular tool in image segmentation is the simplest, yet effective, technique of *thresholding* that relies only on the point or pixel values of the image. Gray-level thresholding segments an image based on the image's value at each point (x, y) relative to a specified threshold value, T. The threshold can be either global or local. A global thresholding technique thresholds the entire image, $f(x, y)$, with a single threshold value, to obtain the thresholded image, $g(x, y)$. On the other hand, a local thresholding technique partitions a given image into sub-images or regions, and thresholds are selected independently based on the local properties of the regions. In either case, the result of thresholding is a binary image containing the pixels or regions that satisfy the thresholding criterion. Thresholding can also be categorized as a point-based or region-based operator [68].

In the simplest case of thresholding, commonly known as *binarization*, a single threshold is specified, and the result is defined as [62]

$$g(x, y) = \begin{cases} 1, & \text{if } f(x, y) > T, \\ 0, & \text{if } f(x, y) \leq T, \end{cases} \tag{2.3}$$

where $T \in G$; here, G is the set of available values for gray levels.

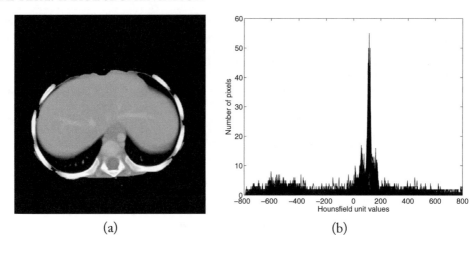

(a)   (b)

Figure 2.1: (a) A 512 × 512-pixel CT image slice, shown with the display range [-400, 400] HU out of [-3024, 1071] HU. Only one cross-section is shown from an exam with 75 slices. (b) Histogram of the image shown in (a), displayed only for the range [-800, 800] HU.

Thresholding requires knowledge of the expected gray levels of the objects of interest and the background, in order to be effective. By considering the image's histogram along with the *prior* knowledge about the expected gray levels of an object of interest, it may be possible to improve the result of thresholding.

A simple result of thresholding the image in Figure 2.1 (a) at +200 HU is displayed in part (a) of Figure 2.2. Notice that most of the bone structures have been identified in the result; however, some parts of the bone structure are not included because they have lower HU values than the threshold.

In most images, several objects of interest or parts thereof may possess different values, and hence require the use of multiple thresholds to perform their segmentation: this approach is known as *multi-thresholding*. In such a case, a set of thresholds, $T = \{T_0, T_1, T_2, ..., T_k\}$, is defined such that all elements in the image satisfying $f(x, y) \in [T_{i-1}, T_i)$, $i = 1, 2, ..., k$, constitute the ith segmented region.

A simple case of multi-thresholding, using $T_0 = 0$ HU and $T_1 = 200$ HU, is demonstrated in Figure 2.2 (b). The histogram of the image, shown in Figure 2.1 (b), has the highest concentration of pixels in the range of [0, 200] HU, which corresponds mostly to the liver and the spleen in the image in Figure 2.1 (a). Note that other tissues having HU values within the specified range are also included in the result of thresholding.

The selection of appropriate thresholds for an application is a difficult task. There exist several techniques to select optimal thresholding parameters, depending on the maximization or minimization of a merit or performance function. Many algorithms rely on parametric and non-parametric

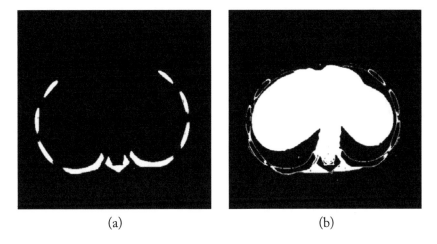

(a)   (b)

Figure 2.2: (a) A simple thresholded version of the image shown in Figure 2.1 (a) using $T = 200$ HU. (b) Multi-thresholded version of the same image using $T_0 = 0$ HU and $T_1 = 200$ HU. The thresholds were determined by using the histogram in Figure 2.1 (b).

statistics [69] to estimate the appropriate threshold. Otsu's method [70] partitions the given image based on discriminant analysis; the procedure calculates an appropriate threshold by maximizing the between-class variance for two possible classes in the image. Other methods, such as those proposed by Pun [71], Kapur et al. [72], and Wong and Sahoo [73], rely on maximizing different measures of entropy in the image.

In order to segment the objects in an image successfully, a minimum separation between the values of the objects of interest and the background is required. Even if the histogram of the image is bimodal, correct segmentation may not be achieved if the objects are located in a background having diverse values [64]. In the case of CT images, several organs within the body have similar characteristics (see Table 1.1), and thresholding may fail to identify the individual structures.

2.4   REGION-BASED METHODS

Segmentation involves partitioning of an image into regions; in this context, thresholding methods provide a global assessment of the potential object candidates in the given image. Thresholding methods identify all potential objects or pixels in the image that satisfy the specified criteria, based solely upon pixel values and neglecting all the spatial information in the image [5]. Region-based segmentation takes the spatial information into consideration, based on the postulate that neighboring pixels within a region have similar values [5]. Such methods aim to partition the given image into distinct and disjoint regions that correspond, ideally, to objects in the image that have spatially connected components [62, 64].

The region-based approach can be described as follows [5, 62]: if R represents the entire image region and segmentation results in n subregions, R_1, R_2, ..., R_n, the results should satisfy

1. $\bigcup_{i=1}^{n} R_i = R$,

2. R_i is a connected region, $i = 1, 2, ..., n$,

3. $R_i \bigcap R_j = \emptyset$ for all $i \neq j$,

4. $\mathcal{P}(R_i) = \text{TRUE}$ for all $i = 1, 2, ..., n$,

5. $\mathcal{P}(R_i \bigcup R_j) = \text{FALSE}$ for all $i \neq j$,

where $\mathcal{P}(R_i)$ is a logical predicate defined over the points or pixels in the set R_i, and $\emptyset$ is the null set.

The conditions listed above indicate that the segmentation should be complete (that is, every pixel must be in a region); the resulting regions should comprise elements that are connected in some predefined sense; the regions should be disjoint and homogeneous; and each region should possess distinct characteristics according to the predicate $\mathcal{P}$.

Region-based segmentation algorithms may be divided into two groups: *region growing* and *region splitting and merging*.

2.4.1   REGION GROWING

Region growing is a procedure that groups pixels or subregions into larger regions based on predefined criteria and connectivity. The basic approach is to begin with a set of initial starting points or seed pixels, and to grow a region by appending the neighboring points of the seeds to the region if they have similar properties, such as specific ranges of gray level, texture, or color. The result of the procedure depends on the selection of seed pixel or pixels, the specification of the inclusion or similarity criteria, and the formulation of the stopping rule.

The specification of the similarity criteria depends not only on the problem under consideration but also on the type of image data available. In addition, specified similarity criteria can yield misleading results if the connectivity or adjacency information is not properly used, and if the seed pixels are not properly selected. Another problem lies in the formulation of the stopping rule. Adaptive region growing [5, 74], region growing using an additive or a multiplicative tolerance [5, 74], region growing using fuzzy sets [5], and region growing using shape or other constraints [5, 62] can improve the accuracy of the result of segmentation.

2.4.2   REGION SPLITTING AND MERGING

An alternative approach to region growing is *region splitting and merging*. Instead of using seeds to grow regions, the splitting step is used to subdivide the entire region R successively into smaller and smaller disjoint regions until the homogeneity criterion or predicate is satisfied by each region. The basic splitting approach is applied as follows: if $\mathcal{P}(R_j) = \text{FALSE}$ for a subregion R_j, subdivide

R_j into quadrants successively, and apply the same procedure to each quadrant. This is an iterative procedure that runs until no further changes are made, or until a stopping criterion is reached [5].

Region splitting could result in adjacent regions with identical or similar properties. This issue may be addressed by using a merging step to allow neighboring homogeneous subregions to combine into larger regions. The process can be described as: merge adjacent regions R_i and R_j for which $\mathcal{P}(R_i \bigcup R_j) = \text{TRUE}$. The procedure is completed when no further merging or splitting is possible [62].

2.5 EDGE-BASED TECHNIQUES

Edge detection is an important part of digital image processing. An edge is the oriented boundary between two regions with relatively distinct gray-level properties. Most edge-detection techniques require the computation of a local derivative or difference [62]. First-order derivatives of a digital image are based on various approximations of the 2D gradient. The gradient of an image $f(x, y)$ at the location (x, y) is defined as the vector

$$\nabla \mathbf{f} = \left[\begin{array}{c} G_x \\ G_y \end{array} \right] = \left[\begin{array}{c} \partial f / \partial x \\ \partial f / \partial y \end{array} \right]. \tag{2.4}$$

The gradient vector points in the direction of maximum rate of change of f at the coordinates (x, y). An important quantity in edge detection is the magnitude of the gradient vector, denoted by ∇f, where

$$\nabla f = |\nabla \mathbf{f}| = \left[G_x{}^2 + G_y{}^2 \right]^{0.5}. \tag{2.5}$$

The magnitude of the gradient vector gives the rate of change of $f(x, y)$ per unit distance in the direction of $\nabla \mathbf{f}$. It is a common practice (but not strictly correct) to refer to ∇f as the *gradient* [62]; the direction of the gradient vector is also an important quantity. If $\alpha(x, y)$ represents the direction (angle) of the vector $\nabla \mathbf{f}$ at (x, y),

$$\alpha(x, y) = \tan^{-1} \left(\frac{G_y}{G_x} \right), \tag{2.6}$$

where the angle is measured with respect to the x axis. The direction of an edge at (x, y) is perpendicular to the direction of the gradient vector at that point.

In digital images, the components of the gradient magnitude in the x and y direction are approximated by first difference operations, defined as

$$\begin{array}{rcl} G_x(x, y) & = & f(x, y) - f(x - 1, y), \\ G_y(x, y) & = & f(x, y) - f(x, y - 1). \end{array} \tag{2.7}$$ $$\tag{2.8}$$

An example of application of the gradient magnitude, calculated using the above equations, is displayed in Figure 2.3. It is evident from parts (b) and (d) of the figure that the gradient magnitude is capable of detecting the edges in both binary and gray-scale images.

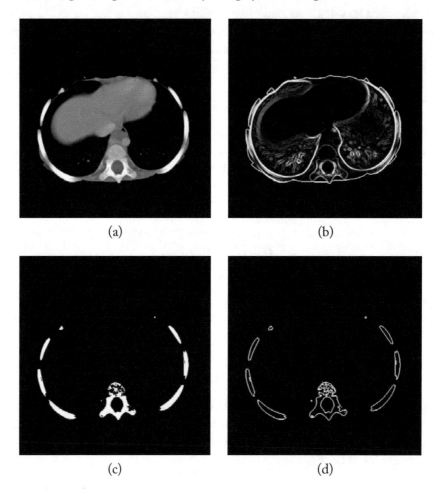

(a)

(b)

(c)

(d)

Figure 2.3: (a) A 512 × 512-pixel CT image showing a section through the chest with the display range HU=[-200, 400] out of [-3024, 1071]. (b) 2D gradient magnitude of the image in (a). Display range is adjusted to [5, 150] HU out of [0, 2371] HU. (c) The image in (a) thresholded within the range [200, 1008] HU. (d) 2D gradient magnitude of the binary image in (c).

To reduce the computational cost, a frequently used approach is to approximate the gradient by the sum of the absolute values of its components, as

$$\nabla f \approx |G_x| + |G_y|, \tag{2.9}$$

which preserves the relative changes in gray levels. The disadvantage of this approach is that the resulting filters will not be isotropic, in general.

In image processing, it is desirable to express the operators in terms of odd-sized masks, centered upon the pixel being processed [5]. The convolution of the mask with the image produces the gradient image. Among the difference-based edge-detection operators, the Prewitt and Sobel operators are simple and popular. These operators provide differencing along with smoothing, which is highly desirable because difference operators enhance noise [62]. The masks can be arranged to detect vertical edges, horizontal edges, and edges oriented at 45° and 135° (See Figure 2.4 and Figure 2.5).

$$
\begin{bmatrix} -1 & -1 & -1 \\ 0 & 0 & 0 \\ 1 & 1 & 1 \end{bmatrix}
\begin{bmatrix} 0 & -1 & -1 \\ 1 & 0 & -1 \\ 1 & 1 & 0 \end{bmatrix}
\begin{bmatrix} -1 & 0 & 1 \\ -1 & 0 & 1 \\ -1 & 0 & 1 \end{bmatrix}
\begin{bmatrix} -1 & -1 & 0 \\ -1 & 0 & 1 \\ 0 & 1 & 1 \end{bmatrix}
$$

Vertical   45°   Horizontal   135°

Figure 2.4: The Prewitt operators.

$$
\begin{bmatrix} -1 & -2 & -1 \\ 0 & 0 & 0 \\ 1 & 2 & 1 \end{bmatrix}
\begin{bmatrix} 0 & -1 & -2 \\ 1 & 0 & -1 \\ 2 & 1 & 0 \end{bmatrix}
\begin{bmatrix} -1 & 0 & 1 \\ -2 & 0 & 2 \\ -1 & 0 & 1 \end{bmatrix}
\begin{bmatrix} -2 & -1 & 0 \\ -1 & 0 & 1 \\ 0 & 1 & 2 \end{bmatrix}
$$

Vertical   45°   Horizontal   135°

Figure 2.5: The Sobel operators.

The Laplacian is a second-order difference operator which is widely used for edge detection. A discrete implementation of the Laplacian is given by the 3×3 convolution kernel

$$
\begin{bmatrix} 0 & 1 & 0 \\ 1 & -4 & 1 \\ 0 & 1 & 0 \end{bmatrix}.
$$

Although the Laplacian has the advantage of being omni-directional, it produces double-edged outputs with positive and negative values at each edge, and is sensitive to noise because there is no averaging included in the operator. The noise sensitivity could be reduced by defining a scalable, smoothing, 2D Gaussian function, where the variance would be the controlling factor of the spatial extent or width of the smoothing function. Combining the Laplacian and the Gaussian, the second-order Laplacian-of-Gaussian (LoG) operator is obtained [5], which is popular and useful for robust omni-directional edge detection.

Canny [75] proposed an approach for edge detection based upon three criteria for good edge detection: multi-directional derivatives, multi-scale analysis, and optimization procedures. Canny suggested that an edge detector should have a low probability of failing to mark real edge points, and a low probability of falsely marking non-edge points, represented by the signal-to-noise ratio (SNR). Furthermore, points marked by the operator should be as close as possible to the center of the true edge: Canny symbolized this aspect by using the reciprocal of the root-mean-squared (RMS) distance of the marked edge from the center of the true edge. Finally, the algorithm should have only one response to a single edge. This was represented by the distances between adjacent maxima in the output. The constraints were numerically optimized to find optimal operators for roof and ridge edges. The analysis was then restricted to consider optimal operators for step edges. Because Canny's method selectively evaluates a directional derivative at each edge, it could produce better results than the LoG filter by avoiding derivatives at other angles that would not be necessary for edge detection and are responsible for increasing the effects of noise. The results of application of the Canny edge-detection method on a gray-scale and a binary image are shown in parts (a) and (b) of Figure 2.6, respectively.

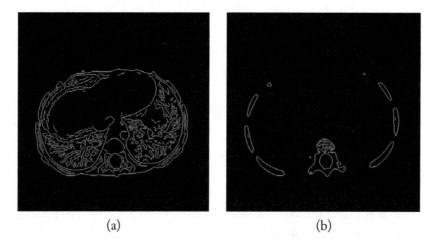

(a)   (b)

Figure 2.6:   Results of application of the Canny edge-detection technique: (a) on the gray-scale image shown in part (a) of Figure 2.3; (b) on the binary image shown in part (c) of Figure 2.3.

The purpose of segmentation is to extract and define complete objects and structures in the given image. However, edge-detection algorithms produce only a series of boundary candidates for the objects. Often, duplicate and/or unconnected edges are produced for a single, true, object boundary. In this context, an *edge-linking* procedure [5] can be used as a post-processing step to link disjoint segments and to obtain complete representations of the boundaries of the ROIs.

2.6   ACTIVE CONTOUR MODELING

Image segmentation is a complicated task due to the large variability in the characteristics of the objects of interest and variation in image quality. In particular, medical images are often corrupted by noise and artifacts, and lack well-defined boundaries or characteristics to identify the ROIs. These factors degrade the results of traditional segmentation methods, such as thresholding, region growing, or edge detection. To address these issues, *active contours* have been extensively studied and widely used in medical image segmentation, with promising results.

Active contours, *snakes*, or *deformable contours* [76] are curves or surfaces defined within an image domain that can move under the influence of internal forces that are defined within the curve or surface itself, and external forces, which are computed from the image data. The internal forces are designed to keep the model smooth during deformation, whereas the external forces are defined to move the model toward an object boundary or other desired feature within the image. There are two basic types of deformable models [76]: *parametric deformable models* and *geometric deformable models*. The technique is being widely studied, with different formulations of the model, forces, and deformation strategies. The method of deformable contours is a fundamental tool in computer vision and image processing, especially with various applications in computer-assisted analysis of medical images [77].

2.6.1   MATHEMATICAL MODEL OF DEFORMABLE CONTOURS

Kass et al. [78] proposed *active contour models* or *snakes* that allow an initial contour to reshape and mold itself to a desired object, relying on constraints based upon the image gradient, derivatives of the contour, and energy functionals measured within the image.

Mathematically, the deformable contour is modeled as a closed curve $\mathbf{C}(s) = (X(s), Y(s))$, for $s \in [0, 1]$. The contour moves through the image domain under the influence of internal forces $\mathbf{F}_{\text{int}}(\mathbf{C})$ and external forces $\mathbf{F}_{\text{ext}}(\mathbf{C})$. The internal energy specifies the tension or smoothness of the contour, whereas the external forces attract the curve toward the object's boundaries.

The contour is modeled as an elastic string subjected to a set of dynamic forces, as follows:

$$\mu \, \frac{\partial^2 \mathbf{C}}{\partial t^2} = \mathbf{F}_{\text{damp}}(\mathbf{C}) + \mathbf{F}_{\text{int}}(\mathbf{C}) + \mathbf{F}_{\text{ext}}(\mathbf{C}) \,, \tag{2.10}$$

where μ is the specific mass of the contour and t refers to time.

The internal force [76] is given by

$$\mathbf{F}_{\text{int}}(\mathbf{C}) = \frac{\partial}{\partial s} \left(\alpha \, \frac{\partial \mathbf{C}}{\partial s} \right) - \frac{\partial^2}{\partial s^2} \left(\beta \, \frac{\partial^2 \mathbf{C}}{\partial s^2} \right) \,, \tag{2.11}$$

where the parameters α and β control the model's tension and rigidity; in practice, they are often chosen to be constants. The first-order derivative discourages stretching and makes the model behave like an elastic string; the second-order derivative discourages bending and makes the model behave like a rigid rod [76].

The damping (viscous) force is given by

$$\mathbf{F}_{\text{damp}}(\mathbf{C}) = -\gamma \, \frac{\partial \mathbf{C}}{\partial t} \, , \tag{2.12}$$

where γ is the viscosity constant; a viscous force is introduced to stabilize the deformable contour around the static equilibrium configuration.

In image segmentation, the mass of the contour is often assumed to be zero, resulting in the following equation:

$$\gamma \, \frac{\partial \mathbf{C}}{\partial t} = \mathbf{F}_{\text{int}}(\mathbf{C}) + \mathbf{F}_{\text{ext}}(\mathbf{C}) \, . \tag{2.13}$$

The external force $\mathbf{F}_{\text{ext}}(\mathbf{C})$ can be either a potential force or a non-potential force, and is related to the characteristics of the image. The external force pushes the contour toward the object's boundary. Several kinds of external forces have been formulated, such as multi-scale Gaussian potential force, pressure force, distance potential force, dynamic distance force, interactive force, and gradient vector flow (GVF).

When the external and the internal forces become equal, the force field attains equilibrium and the contour stabilizes. Discretizing Equations 2.11, 2.12, and 2.13 permits the implementation of the deformable-contour technique as a numerical simulation problem in digital devices. The spatial and time derivatives can be approximated by finite differences.

2.6.2   GRADIENT VECTOR FLOW

A traditional definition of the external force is based on the edge map of the given image $I(x, y)$ as the intensity of the edges in the image, computed as

$$E(x, y) = |\nabla I(x, y)|^2. \tag{2.14}$$

The external force is formulated as $\mathbf{F}_{\text{ext}} = -\nabla E(x, y)$.

This method generates a force field in the vicinity of edges that pushes the deformable contour toward strong edges. Ideally, the deformable contour will be displaced and deformed to match the boundaries of the object being segmented. However, the performance of the deformable contour depends on proper initialization. If the contour is not initialized close to the object's boundary, the external force of the deformable contour will have a low intensity and will not result in proper convergence of the deformable contour to the boundary of the object of interest.

To address this problem, Xu and Prince [79] deployed a vector diffusion equation that diffuses the gradient of an edge map in regions distant from the boundary, resulting in a different force field called the *Gradient Vector Flow* (GVF) field. The GVF field $(u(x, y), v(x, y))$ is defined as the solution of the following system of differential equations:

$$\lambda \nabla^2 u - \left(u - \frac{\partial E}{\partial x}\right) \left[\left(\frac{\partial E}{\partial x}\right)^2 + \left(\frac{\partial E}{\partial y}\right)^2\right] = 0, \tag{2.15}$$

$$\lambda \nabla^2 v - \left(v - \frac{\partial E}{\partial y}\right) \left[\left(\frac{\partial E}{\partial x}\right)^2 + \left(\frac{\partial E}{\partial y}\right)^2\right] = 0, \tag{2.16}$$

where the arguments (x, y) have been omitted for simplicity, and ∇^2 is the Laplacian operator. The parameter λ is a regularization parameter that controls the extent of smoothness in the GVF field.

In the present study, to achieve further uniformity and extent of the external forces, the following mapping function is applied to the gradient magnitude of the edge map:

$$f\left[|\nabla E|\right] = K_2 \left[1 - \exp\left(\frac{-|\nabla E|}{K_1}\right)\right], \tag{2.17}$$

where $|\nabla E|$ represents the gradient magnitude of the edge map, K_1 determines the rate of convergence, and K_2 determines the asymptote of convergence.

A similar mapping function is also used to remap the GVF as

$$f(\mathbf{w}(x, y)) = K_2 \left[1 - \exp\left(-\frac{|\mathbf{w}(x, y)|}{K_1}\right)\right] \frac{\mathbf{w}(x, y)}{|\mathbf{w}(x, y)|}, \tag{2.18}$$

where $\mathbf{w}(x, y) = [u(x, y) \ v(x, y)]^T$ is a vector representing the GVF components. The mapping functions in Equation 2.17 and Equation 2.18 saturate high values at an upper bound, while amplifying small values.

The result of application of a deformable model with GVF is demonstrated in Figure 2.7. The initial contour, shown in green in Figure 2.7 (a), was obtained by using least-squares estimation of the diaphragm, considering the base of the lung as reference (see Section 5.3). The result after processing the contour with the deformable model is shown in red in Figure 2.7 (b). Note that the contour is able to converge to the nearest edges because of close initialization.

Curves that are initialized far from the desired object's boundaries may converge to the nearest, but potentially undesired, edges. Figures 2.7 (c) and (d) demonstrate the effect of initializing far from the actual boundary. The initial contour in green in Figure 2.7 (c) was drawn manually. Notice that the resulting contour, shown in red in Figure 2.7 (d), has converged to boundaries external to the diaphragm.

Because deformable models are implemented in the continuous domain, the resulting boundary can achieve subpixel accuracy, which is a highly desirable property in medical imaging applications. The principles of deformable contours are also applicable to 3D deformable models (that is, deformable surfaces) [76].

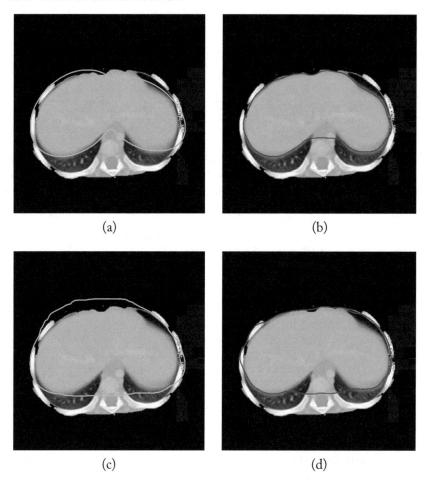

(a)

(b)

(c)

(d)

Figure 2.7:   (a) Original image with the initial contour shown in green. (b) Result of applying the deformable contour model and GVF, shown in red. (c) Original image with the initial contour shown in green, initialized away from the desired boundaries of the diaphragm. (d) Result of applying the deformable contour model and GVF, shown in red.

2.7   THE HOUGH TRANSFORM

The *Hough transform* [80, 81] is a useful tool to detect any shape that can be represented by a parametric function (for example, a straight line, a circle, a sphere, an ellipse, an ellipsoid, or a more complex shape), using the information related to edges in the form of a binary image. The major strength of the Hough transform is the ability to recognize shapes and object boundaries, even with a sparse edge map.

To detect straight lines in images, Hough [80] proposed to represent straight lines in the image function $h(x, y)$ using the slope-intercept equation, $y = mx + c$, where m is the slope and c is the y-intercept. In the parameter space, or the Hough domain or space (m, c), any straight line in the image space is represented by a single point. Two points that are collinear in the image and lie on the line, $y = m_0 x + c_0$, will coincide in the parameter space at the point (m_0, c_0). The procedure to detect shapes requires that the parameter space be discretized into accumulator bins. The parameters are computed for every pixel that has been marked as a boundary pixel, and the corresponding accumulator bins are incremented. The final Hough space may be thresholded at a predefined level to detect lines above or below a specified length.

Because both the slope m and the intercept c have unbounded ranges, practical difficulties arise in their computation and representation. To overcome these limitations, Duda and Hart [81] proposed an alternative form of parameterization using the normal parameters (ρ, θ). A straight line can be specified by the angle θ of its normal and its distance ρ from the origin by the equation

$$\rho = x \cos \theta + y \sin \theta. \tag{2.19}$$

In this representation, the main advantage is that θ is restricted to $[0, 180°]$ or $[0, 360°]$, and ρ is limited by the size of the image. The origin may be chosen at the center of the image or at any other arbitrary point. The value of ρ can be considered to be negative for normals to lines extending below the horizontal axis $x = 0$ in the image, with the origin at the center of the image and the range of θ defined to be $[0, 180°]$. It is also possible to maintain ρ to be positive, with the range of θ extended to $[0, 360°]$.

Figure 2.8 (a) shows an image with two straight lines with $(\rho, \theta) = (-100, 30°)$ and $(\rho, \theta) = (20, 60°)$. Figure 2.8 (b) shows the parameter space (ρ, θ), and demonstrates the anticipated sinusoidal patterns. Maxima occur at the expected intersection locations of $(\rho, \theta) = (-100, 30°)$ and $(\rho, \theta) = (20, 60°)$. The origin was considered to be at the center of the image. Note that the shorter line in the image corresponds to a lower intensity point in the Hough space than the longer line. Detection of local maxima in the Hough space could lead to the detection of the corresponding straight lines in the original image.

The Hough transform can also be extended to the detection of complex parametric curves [81]. For example, all points along the circumference of a circle of radius c centered at $(x, y) = (a, b)$ of an image $h(x, y)$ satisfy the relationship

$$(x - a)^2 + (y - b)^2 = c^2. \tag{2.20}$$

A circle is represented by a single point in the 3D parameter space (a, b, c). The points along the circumference or the edge of a circle in the (x, y) plane describe a right circular cone in the Hough parameter space, which is limited by the image size. Prior knowledge of the possible radius values could assist in fast and accurate detection by overcoming the effects of artifacts. Figure 2.9 (a) shows a circle of radius 25 pixels, centered at $(a, b) = (50, 50)$. Figures 2.9 (b) through (f) show the Hough parameter space for all (a, b) in the range $[1, 100]$, and $c = 20, 24, 25, 26$, and 30 pixels,

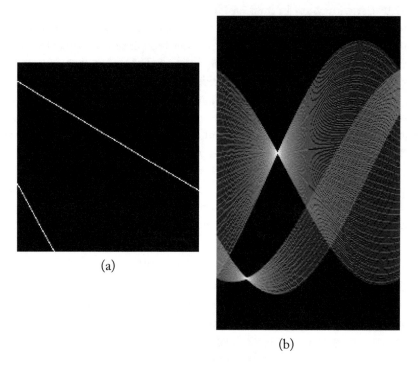

(a)

(b)

Figure 2.8: (a) An image with two straight lines with $(\rho, \theta) = (-100, 30°)$ and $(\rho, \theta) = (20, 60°)$. The limits of the x and y axes are ± 100. (b) The Hough transform parameter space for the image in (a). The display intensity is $\log_{10}(1 + accumulator\ cell\ value)$. The vertical axis represents $\rho = [-150, 150]$ and the horizontal axis represents $\theta = [0°, 180°]$.

respectively. There is a clear peak at $(a, b, c) = (50, 50, 25)$, shown in Figure 2.9 (d), indicating the presence of a circle of radius $c = 25$ at $(x, y) = (50, 50)$.

The use of the Hough transform to detect seed voxels in the spinal canal is described in Section 4.5.

2.8   THE CONVEX HULL

Decomposing a boundary into segments can be useful in reducing the boundary's complexity and simplifying its description. This approach is particularly useful when the boundary contains one or more significant concavities, with the associated shape information. However, a boundary with multiple disjoint segments may create difficulties in further analysis.

The *convex hull* of a set of points is the smallest convex set that contains the points. The convex hull is a fundamental construction in computational geometry [82]. The problem of finding convex

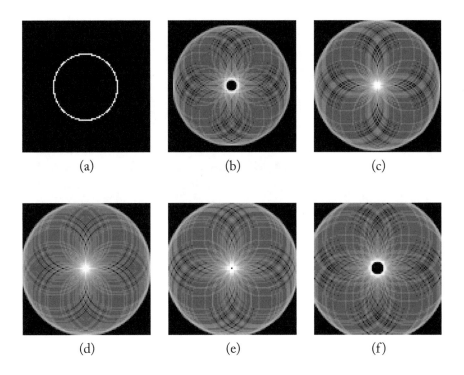

Figure 2.9: (a) A 100×100-pixel image with a circle of radius $c = 25$ pixels, centered at $(a, b) = (50, 50)$. The image represents the range $[1, 100]$ for x and y, with $(x, y) = (1, 1)$ at the top-left corner. Images (b) through (f) represent the Hough parameter space (a, b, c) of the image in (a). Each image is of size 100×100 pixels, and represents the range of the parameters a and b for a given c. Hough parameter space for (b) $c = 20$ pixels, (c) $c = 24$ pixels, (d) $c = 25$ pixels, (e) $c = 26$ pixels, and (f) $c = 30$ pixels. The maximum value in the Hough parameter space occurs at the center $(a, b) = (50, 50)$ for $c = 25$, as expected.

hulls has practical applications in image processing, pattern recognition, statistics, and geographic information systems (GIS).

Mathematically, a set A is said to be *convex* if the straight-line segment joining any two points in A lies entirely within A. The convex hull H of an arbitrary set S is the smallest convex set containing S. The set difference $H - S$ is called the *convex deficiency* of S. The concepts of a convex hull and its deficiency are useful in describing an entire region as well as its boundary.

In part (b) of Figure 2.10, an example of the convex hull is shown: the actual region is shown in gray; the rest of the region belonging to the convex hull is the convex deficiency, and is shown in white.

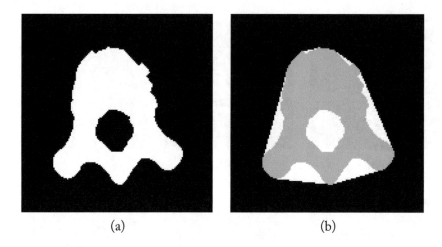

(a)   (b)

Figure 2.10: (a) A 150×150-pixel image containing a binarized image of a slice through vertebral column. (b) The convex hull of the image in part (a). The actual region is shown in gray and the convex deficiency is shown in white.

2.9   FUZZY SEGMENTATION

Many traditional segmentation algorithms, such as thresholding and region growing, employ the *all-or-nothing* assignment technique, and partition an image in a "crisp" manner. This approach is effective when the objects in the image are clearly delineated or defined. To accommodate for the heterogeneity of organs, the inherent variability of biological systems, and overlapping ROIs in medical images, the rigid parameters of such algorithms need to be made flexible. The computational representation of imprecise, vague, and ambiguous parameters are possible with the advent of segmentation techniques based on *fuzzy sets*.

Fuzzy sets were introduced by Zadeh [83] as a new way to represent the vagueness or imprecision encountered in everyday life. Fuzzy sets originate from the generalization of conventional set theory, and serve to quantify the imprecision in information to varying degrees [84, 85].

In the context of image segmentation, the notion of fuzzy sets combined with the concept of fuzzy connectivity, as proposed by Rosenfeld [86], is a powerful tool to quantify not only the degree to which the elements of an image satisfy the properties of the object of interest individually, but also to quantify the way in which they are related: this forms the basis for fuzzy segmentation [85].

2.9.1   FUZZY SETS

Theoretically, a fuzzy set $\mathcal{A}$ in a reference set $\mathcal{X}$ can be characterized by a membership function, $m_{\mathcal{A}}$, that maps all elements in $\mathcal{X}$ into the interval [0, 1]. The fuzzy set may be represented as a set of fuzzy pairings, written as

$$\mathcal{A} = \{(x, m_{\mathcal{A}}(x))|x \in \mathcal{X}\}, \tag{2.21}$$

where $m_{\mathcal{A}}$ is the membership or characteristic function, defined as

$$m_{\mathcal{A}}(x) = x \mapsto [0, 1], \text{ for } x \in \mathcal{X}. \tag{2.22}$$

The membership value $m_{\mathcal{A}}(x)$ denotes the degree to which an element x satisfies the properties of the set $\mathcal{A}$. Values close to unity represent high degrees of membership, whereas values near zero represent the lack of similarity with the characteristics of the set.

In order to manipulate fuzzy sets, the foundational set operators can be extended to handle varying degrees in the relationship. Given two fuzzy sets $\mathcal{A}$ and $\mathcal{B}$ with the membership functions $m_{\mathcal{A}}(x)$ and $m_{\mathcal{B}}(x)$, the standard set-theoretic relations and operations are defined ($\forall x \in \mathcal{X}$) as:

Equality (=) $\mathcal{A} = \mathcal{B} \Leftrightarrow m_{\mathcal{A}} = m_{\mathcal{B}},$

Containment ($\subset$) $\mathcal{A} \subset \mathcal{B} \Leftrightarrow m_{\mathcal{A}} \leq m_{\mathcal{B}},$

Complement ($\sim$) $m_{\tilde{\mathcal{A}}}(x) = 1 - m_{\mathcal{A}}(x),$

Intersection ($\cap$) $m_{\mathcal{A} \cap \mathcal{B}}(x) = \min\{m_{\mathcal{A}}(x), m_{\mathcal{B}}(x)\},$

Union ($\cup$) $m_{\mathcal{A} \cup \mathcal{B}}(x) = \max\{m_{\mathcal{A}}(x), m_{\mathcal{B}}(x)\}.$

2.9.2   FUZZY MAPPING

As described in Section 2.9.1, a fuzzy set $\mathcal{A}$ is represented by a membership function $m_{\mathcal{A}}$, which maps numbers into the entire unit interval [0, 1]. The value $m_{\mathcal{A}}(r)$ is called the *grade of membership* of r in $\mathcal{A}$: this function indicates the degree to which r satisfies the membership criteria defining $\mathcal{A}$.

For example, consider the set of numbers that are "close to 10". For crisp sets, the concept of "closeness" has no quantitative definition. However, it is possible to quantify this relationship by using a membership function. In defining $m_{\mathcal{A}}(r)$ for this example, three properties should be satisfied [84]:

1. Normality: $m_{\mathcal{A}}(10) = 1$.

2. Monotonicity: The closer the value of r is to 10, the closer $m_{\mathcal{A}}$ should be to 1.

3. Symmetry: Numbers equally far from 10, such as 9 and 11, should have equal membership.

The unnormalized Gaussian function, defined as

$$m_{\mathcal{A}}(r) = \exp\left\{-\frac{(r-\mu)^2}{2\sigma^2}\right\},$$

(2.23)

is an example of a fuzzy membership function satisfying the properties listed above; the function is shown in Figure 2.11. The parameters μ and σ characterize the set $\mathcal{A}$; for the present example, $\mu = 10$, and σ may be chosen depending upon the desired rate of decrease in the membership function for values away from 10. Fuzzy membership functions should, in general, satisfy the properties listed above.

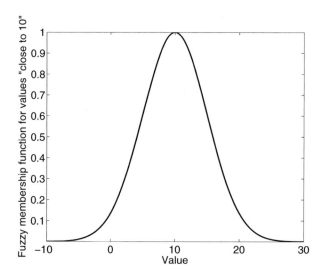

Figure 2.11: Unnormalized Gaussian function to represent values "close to 10".

In image processing, fuzzy sets could be used to quantify the degrees of similarity of image elements to the objects of interest. For example, using the membership function in Equation 2.23 with the parameters $\mu_b = +412$ HU and $\sigma_b = 156$ HU for bone, $\mu_l = -528$ HU and $\sigma_l = 121$ HU for lung tissue, and $\mu_s = +18$ HU and $\sigma_s = 14$ HU for the spinal canal, information related to the bone structures, lung tissues, or the spinal canal can be enhanced to assist in segmentation. For the image in Figure 2.12 (a), the results shown in Figures 2.12 (b), (c), and (d), represent the desired bone structures, lung tissues, or the spinal canal as bright regions with high membership values. In each case, the undesired structures appear faint, possessing lower degrees of similarity or membership. There are areas with intermediate shades of gray in the resulting images: these denote areas where doubt exists in regard to satisfying the characteristics of the object of interest. Ambiguity arises when other regions possess the same characteristics as the object of interest, visible significantly in part (d) of Figure 2.12 for the spinal canal.

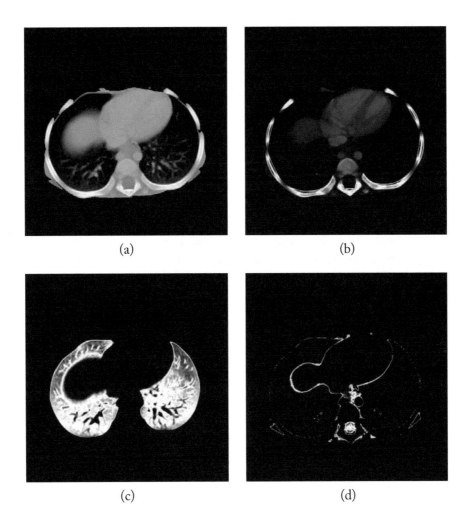

(a)   (b)

(c)   (d)

Figure 2.12: (a) A 512 × 512-pixel CT image containing a section through the chest. The slice is shown after contrast adjustment for improved viewing. (b) Fuzzy mapping of the image in (a) to identify bone structure. The parameters for bone were estimated as $\mu_b = +412$ HU and $\sigma_b = 156$ HU. (c) Fuzzy mapping of the image in (a) to identify the lung tissues. The parameters used are $\mu_l = -528$ HU and $\sigma_l = 121$ HU. (d) Fuzzy mapping of the image in (a) to identify the spinal canal. The parameters were estimated to be $\mu_s = +18$ HU and $\sigma_s = 14$ HU. In each case, Equation 2.23 was used to obtain the fuzzy membership value.

Fuzzy sets serve as an effective and flexible way to represent information with imprecise characteristics. As seen in the examples in Figure 2.12, fuzzy membership can identify not only the regions that most likely belong to the desired objects, but can also show regions where some doubt exists in categorizing the objects. The membership function operates globally, and identifies all elements in the image that demonstrate characteristics similar to those of the object of interest. There could be several potential candidates for the desired object, and the membership image can be modified to evaluate the manner of grouping the candidates or image elements to extract specific objects. For this purpose, fuzzy connectivity may be employed for efficient identification and segmentation, as described in the next section.

2.9.3   FUZZY CONNECTIVITY

Fuzzy connectivity [85, 86] is a popular and powerful tool to perform segmentation of medical images. The *degree of connectedness* between a given set of points in an image is an indication of the extent to which the points comprise the same object. For a given object, there should exist a high level of similarity between the corresponding image elements, and direct links connecting all image elements belonging to the same object. The links as above are known as *paths* that connect image elements via their neighboring elements.

The degree of connectedness, $m_{\mathcal{A}}$, between two arbitrary points, p and q, in the fuzzy domain for a set $\mathcal{A}$, is given by

$$C_{m_{\mathcal{A}}}(p, q) = \max_{L_{p,q}} \left[\min_{1 \leq i \leq n} m_{\mathcal{A}}(p_i) \right], \tag{2.24}$$

where $L_{p,q}$ denotes the set of all possible paths from the point p to the point q, p_i is a point along the path from $p_1 = p$ to $p_n = q$, and $m_{\mathcal{A}}$ is a fuzzy membership function. In other words, the connectedness of two pixels (or voxels) p and q is dependent on all possible connecting paths between the two pixels (or voxels). A connecting path is formed from a sequence of links between successive adjacent pixels (or voxels) in the path. The strength of each link is the affinity between the adjacent voxels in the link, and the strength of a path is the strength of its weakest link. Equation 2.24 can be interpreted as the strongest of all possible paths connecting the point p with the point q, where each path is assigned the value of the weakest link along the path.

For a given object, the points belonging to the object should possess a high degree of connectedness due to the strong resemblance based on the fuzzy membership and the existence of strong paths connecting them. Background or undesired elements should possess a low degree of connectedness with the object: although there could exist paths between the desired and undesired elements, they would be expected to possess low membership values.

Morphological reconstruction, discussed in Section 2.10.3, operates on the notion of a connection cost, or the minimum distance between specific points in a defined set. Because of the similarity between fuzzy connectivity and morphological reconstruction, the opening-by-reconstruction procedure may be deployed in conjunction with fuzzy mapping to achieve segmentation using fuzzy connectivity.

2.10 MORPHOLOGICAL IMAGE PROCESSING

Mathematical morphology [65, 87] refers to a branch of nonlinear image processing that concentrates on the analysis of geometrical structures within an image. It is based on conventional set theory and serves as a powerful tool for image processing and analysis. In addition to image segmentation, mathematical morphology provides methods for image enhancement, restoration, edge detection, texture analysis, feature generation, compression, skeletonization, component analysis, curve filling, thinning, and shape analysis [65].

Morphological image processing is based on geometry, and is used to analyze the effects of the application of a specified geometric form known as the *structuring element* to the given image. The aim is to probe the image with the structuring element and quantify the manner in which the structuring element fits, or does not fit, within the image. The type of information produced by the application depends upon the shape and size of the structuring element [88], and the nature of the information in the image.

The characterization of fitting the structuring element to the image depends on the basic Euclidean-space *translation* operation. Considering an image, A, in $\mathbb{R}^2$, the translation of A by a point x is defined as

$$A + x = \{a + x | a \in A\}, \tag{2.25}$$

where $+$ refers to vector addition. The nature of probing is to mark the positions (translations) of the structuring element where it fits in the image.

The most elementary set operations relating to mathematical morphology should have two desirable properties: they should be *increasing* and *translation invariant* [87, 88]. A set operator Ψ is *increasing* if

$$F_1 \subseteq F_2 \Rightarrow \Psi(F_1) \subseteq \Psi(F_2), \tag{2.26}$$

where F_1 and F_2 are subsets of $\mathbb{R}^2$. This property ensures that order is preserved; an increasing operator Ψ forbids an object F_1 that is occluded by an other object F_2 to become visible after processing. In other words, if F_1 is a subset of F_2, then their outputs after processing will also be subsets with the same order.

Translation invariance is another property that plays an important role: it ensures that the various objects within an image are processed the same way no matter where they are located in the image plane. The operator Ψ is considered to be translation invariant if

$$\Psi(F + h) = \Psi(F) + h, \tag{2.27}$$

where h denotes every possible translation in $\mathbb{R}^2$. An operator may be increasing but not translation invariant, or vice versa. Most frequently, operators that are both increasing and translation invariant are used in morphological image processing.

The fundamental morphological image processing operations are based on *Minkowski algebra* [88, 89]. Mathematical morphology-based image processing was originally developed to process binary images; however, the theory has been extended to include gray-scale images, as discussed in Section 2.10.2.

2.10.1 BINARY MORPHOLOGICAL IMAGE PROCESSING

The fundamental operation of mathematical morphology is *erosion*. The translation-invariant erosion operation is known as *Minkowski subtraction* in set theory [88, 89], and is defined as

$$
\begin{aligned}
F \ominus B &= \{h \in \mathbb{R}^n | (B + h) \subseteq F\} \\
&= \bigcap_{b \in B} F - b ,
\end{aligned}
\tag{2.28}
$$

where F and B are subsets of $\mathbb{R}^n$, B is the structuring element for the purpose of eroding F, and h is an element of the set of all possible translations. Note that for a digitized image, F and B are subsets of $\mathbb{Z}^n$ and $h \in \mathbb{Z}^n$. The erosion of F by B ($F \ominus B$) comprises all points h in $\mathbb{R}^n$ such that the translated structuring element $B + h$ fits entirely inside F. In terms of set theory, $F \ominus B$ is formed by translating F by every element in B, and taking the intersection, $\bigcap$, of the results so obtained. If the origin lies within the structuring element, then erosion has the effect of shrinking, and the eroded image is a subset of the original image. Protrusions smaller than the structuring element are eliminated.

On the other hand, translation-invariant *dilation* is known as Minkowski addition [88, 89], and is defined as

$$
\begin{aligned}
F \oplus B &= \left\{h \in \mathbb{R}^n | (\check{B} + h) \bigcap F \neq \emptyset\right\} \\
&= \bigcup_{b \in B} F + b ,
\end{aligned}
\tag{2.29}
$$

where $\check{B} = \{-b | b \in B\}$ is the reflection of B with respect to the origin. The dilation of F by B, ($F \oplus B$), comprises all points h such that the reflected structuring element, $\check{B}$, translated by h, intersects F. In terms of set theory, a dilation is the union, $\bigcup$, of all copies of F translated by every element in B. Because, dilation involves a fitting into the complement of an image, it represents filtering on the outside, whereas erosion represents filtering on the inside. If the structuring element includes the origin, dilation fills in small holes (relative to the structuring element) and intrusions into the image.

In practical implementation of erosion, if any of the pixels within the neighborhood defined by the structuring element is 'off' (that is, set to 0), the output pixel is also set to 0. On the other hand, in the case of dilation, if any of the pixels within the neighborhood defined by the structuring element is 'on' (that is, set to 1), the output pixel is also set to 1.

The basic morphological image processing operations are demonstrated in Figure 2.13, where the original image is shown in part (a) of the figure. B is a disk-shaped, flat, structuring element of radius 5 pixels shown as a yellow disk at the top-left corner of the image in part (a) of Figure 2.13; the neighborhood defined by the structuring element is shown in part (b) of the figure. In the case of erosion, shown in part (c) of Figure 2.13, the operation has the effect of "shrinking" the original object according to the structuring element, and has removed parts smaller than the structuring element. For example, the regions marked by the circles in red, green, cyan, orange, and yellow in the original image have been totally eliminated, leaving disjoint and disconnected structures. Small holes have been made bigger. The holes within the corresponding regions marked by the circles are responsible for the removal of the small objects in the image.

On the other hand, the dilation operation has expanded the original object and filled in most of the holes in the image; see part (d) of Figure 2.13. Observe that the holes marked by the circles in red and violet are totally filled in, whereas the holes marked by the green and cyan circles are not totally filled because the dilation operation is not able to fill in the holes that are larger than the structuring element.

In addition to erosion and dilation, there are two secondary morphological image processing operations that play key roles in image processing: *morphological opening* and *morphological closing* [88]; both possess the property of *idempotency*, that is, consecutive application of the same operator will not have any effect. This highly desirable property ensures that the algorithm stops once stability has been reached. Both operators are composed by the alternate application of the elementary morphological operations.

Morphological opening is achieved by applying a translation-invariant erosion followed by a translation-invariant dilation, as

$$F \circ B = (F \ominus B) \oplus B. \tag{2.30}$$

Opening has the effect of removing objects or details smaller than the structuring element, B, while smoothing the edges of the remaining objects, as shown in Figure 2.14 (a). It also disconnects objects connected by branches that are smaller than the structuring element. The image in part (a) of Figure 2.14 can be interpreted as the dilation of the image in part (c) of Figure 2.13. Note that the opening operation has disconnected the connected parts or removed some parts in the original image (see the regions marked by the circles in red, green, yellow, and cyan).

The dual of opening is known as *morphological closing*, which is achieved by the dilation of F with B, followed by an erosion with B, and is defined as

$$F \bullet B = (F \oplus B) \ominus B. \tag{2.31}$$

Closing has the effect of filling in holes and intrusions that are smaller than the structuring element, as shown in Figure 2.14 (b). The image in part (b) of Figure 2.14 can be interpreted as the erosion of the image in part (d) of Figure 2.13. The same disk-type flat structuring element of radius 5 pixels, as shown in Figure 2.13 (a), was used for all the operations described above.

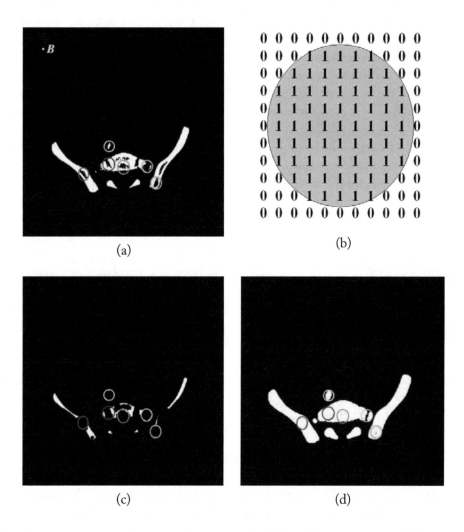

(a)

(b)

(c)

(d)

Figure 2.13: (a) Original image of 512×512 pixels with different regions marked by colored circles to demonstrate the application of binary morphological image processing operations. The disk-shaped flat structuring element B, of radius 5 pixels, is shown in yellow at the top-left corner. (b) The neighborhood defined by the structuring element. (c) Result of application of erosion. The regions marked by the circles indicate removal of parts from the original image and creation of disjoint regions: see the region marked by the yellow circle in part (a). (d) Result of application of dilation. The regions marked by the circles indicate the filling of holes as well as connecting disjoint objects. The regions marked by the green and cyan circles have not been filled in because the structuring element is smaller than the regions.

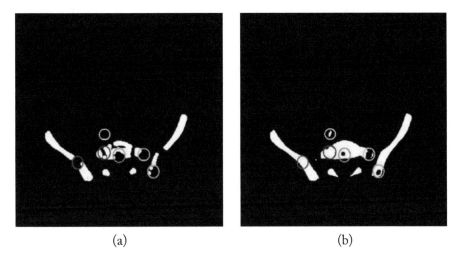

(a)   (b)

Figure 2.14: (a) Result of application of morphological opening to the image shown in part (a) of Figure 2.13 using the same structuring element. The regions marked by the circles indicate the removal of parts from the original image and creation of disconnected regions. (b) Result of application of morphological closing. The regions marked by the circles in red and violet are completely filled in, whereas the regions marked by the other circles are partially filled.

The effects of applying a morphological image processing operator to an image are strongly dependent on the shape and size of the structuring element. The results obtained with the use of different types or sizes of structuring element will be noticeably different.

Morphological thinning [62] can be defined in terms of the hit-or-miss transform [62, 88]. The thinning of A by a structuring element B is denoted by $A \otimes B$. The symmetrical thinning of A is based on a sequence of structuring elements

$$\{B\} = \{B^1, B^2, B^3, \ldots, B^n\}, \tag{2.32}$$

where B^i is a rotated version of B^{i-1}. Thinning with a sequence of structuring elements can be defined as

$$A \otimes \{B\} = ((\ldots((A \otimes B^1) \otimes B^2)\ldots) \otimes B^n). \tag{2.33}$$

The process is repeated until no further changes occur. Thinning removes pixels from the outer sides of an object with no holes; for an object with holes, it removes pixels from the inner and the outer sides. The thinning operation can iteratively remove pixels until convergence is achieved, so that an object without holes shrinks to a minimally connected stroke, and an object with holes shrinks to a connected ring halfway between each hole and the outer boundary.

A skeleton is a representation of a binary image that summarizes the associated object's shape, and conveys useful information about its size, orientation, and connectivity [88]; it is also known as the *medial axis* [90, 91]. Many algorithms have been developed to obtain the skeleton [92]. The skeleton of A can be expressed in terms of erosions and openings [62, 93], as

$$S(A) = \bigcup_{k=0}^{K} S_k(A) , \tag{2.34}$$

where,

$$S_k(A) = (A \ominus kB) - (A \ominus kB) \circ B. \tag{2.35}$$

Here, B is the structuring element, and $(A \ominus kB)$ indicates k successive erosions of A by B; K is the last iterative step before A erodes to an empty set. In practical application, morphological skeletonization removes pixels on the boundaries of objects but does not allow objects to break apart; the pixels remaining make up the skeleton when the algorithm converges. Because the process is invertible, it is possible to reconstruct the original object from its skeleton [88, 89].

Examples of the application of morphological thinning and skeletonization are displayed in Figures 2.15 and 2.16. In Figure 2.15, a simple image is used to illustrate the use of morphological thinning and skeletonization, and results are shown after several iterations. In part (d) and (f) of the same figure, the results are shown after convergence. If the object has holes and complicated structure, morphological thinning and morphological skeletonization produce results that are not simple. Results of application of morphological thinning and skeletonization to the image shown in part (a) of Figure 2.13 are displayed in Figure 2.16.

Binary morphological operators provide a good foundation on which methods may be developed to process different types of objects in gray-scale images.

2.10.2 GRAY-SCALE MORPHOLOGICAL IMAGE PROCESSING

In order to extend mathematical morphology to gray-scale images, it is necessary to define operators to combine gray-scale images in a way that is compatible with union and intersection, which form the basis of binary morphological image processing. A binary image is a special case of a gray-scale image (that is, the number of gray levels is two). The technique known as *threshold decomposition*, which decomposes a gray-scale image into a series of "stacked" binary images, can be used to show that the *supremum* and *infimum* operations, respectively, are gray-scale analogs of set-theoretic union and intersection operations [88].

Given two images g and h on the same image domain D, the point-wise supremum $\bigvee(g(x, y), h(x, y))$ is the maximum (the larger value) of the two images for the corresponding (x, y). On the other hand, the point-wise infimum $\bigwedge(g(x, y), h(x, y))$ is the minimum (the lower value) of the two images for the corresponding points. Using these analogs and techniques, it is possible to extend the basic morphological image processing operators to gray-scale images.

For an image, $f(x, y)$, the translation-invariant gray-scale erosion operation $(f \ominus b)(x, y)$ is defined as

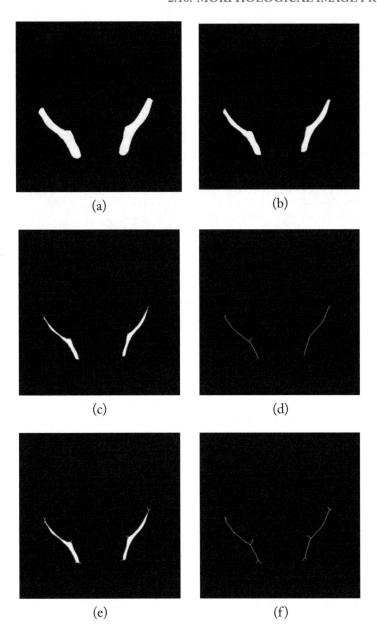

(a)

(b)

(c)

(d)

(e)

(f)

Figure 2.15: (a) A 512×512-pixel CT slice to illustrate the application of morphological thinning and morphological skeletonization. Result of application of morphological thinning: (b) after five iterations, (c) after 10 iterations, and (d) after 20 iterations (convergence). Result of application of morphological skeletonization: (e) after 10 iterations and (f) after 21 iterations convergence).

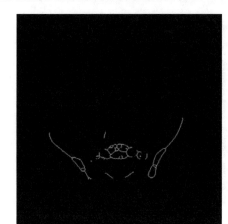 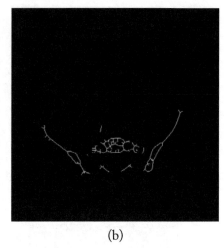

(a)   (b)

Figure 2.16: (a) Result after 14 iterations (convergence) of application of morphological thinning to the image shown in part (a) of Figure 2.13. (b) Result after 13 iterations (convergence) of application of morphological skeletonization on the same figure.

$$(f \ominus b)(x, y) = \bigwedge_{\alpha, \beta \in B} f(x + \alpha, y + \beta), \tag{2.36}$$

where $b(x, y)$ is a flat structuring element defined as

$$b(x, y) = \begin{cases} 0 & \text{for } (x, y) \in B, \\ -\infty & \text{otherwise.} \end{cases} \tag{2.37}$$

The flat erosion replaces the value of an image f at a pixel (x, y) by the infimum of the values of f over the structuring element B.

For an image, $f(x, y)$, the gray-scale dilation operation $(f \oplus b)(x, y)$ is defined as

$$(f \oplus b)(x, y) = \bigvee_{\alpha, \beta \in B} f(x - \alpha, y - \beta), \tag{2.38}$$

where $b(x, y)$ is a flat structuring element defined as in Equation 2.37.

The flat dilation replaces the value of an image f at a pixel (x, y) by the supremum of the values of f over the reflected structuring element $\breve{B}$.

Similar to the binary case, gray-scale opening and closing are defined by combining the basic gray-scale operators, erosion and dilation, as previously defined.

The basic morphological image processing operations provide potential solutions to problems involving edge detection, shape analysis, and image enhancement. In addition, mathematical morphology provides means for the evaluation of the connectivity of regions in an image based on gray-scale values and an initial region (seed or marker), and is a powerful tool for extraction of the specific objects from binary and gray-scale images.

2.10.3 MORPHOLOGICAL RECONSTRUCTION

Morphological *opening-by-reconstruction* is a powerful tool to evaluate the connectivity of objects in images [94]. Opening-by-reconstruction (hereafter referred to simply as "reconstruction") is an iterative procedure that can extract ROIs from an image identified or selected by a set of *markers* [69, 88]. *Markers*, commonly known as *seed pixels*, form the initial point to start the process, and must be contained within the mask of the objects of interest. In the case of binary images, reconstruction is the process to perform extraction or segmentation of the objects connected to the markers in the image, as shown in Figure 2.17.

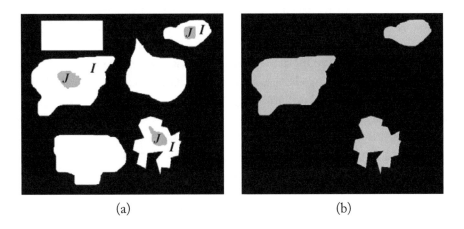

(a)   (b)

Figure 2.17: Schematic illustration of reconstruction with a binary image. (a) The gray objects represent the markers (J) and the white regions are the masks (I). Note that the markers must be totally contained within the mask. (b) The reconstruction of the selected white regions (masks) based on the gray markers.

To define binary reconstruction, let I and J be defined on the image domain D such that $J \subseteq I$. The reconstruction of the image I (called the mask) by the image J (called the marker or seed) can be performed by iterating the elementary geodesic dilation operation on J until further iterations result in no net change. An elementary geodesic dilation, $\delta^{(1)}$, of the marker, J, is simply a standard dilation, $\oplus$, with a unit-sized structuring element, B, followed by an intersection with the mask, I:

$$\delta_I^{(1)}(J) = (J \oplus B) \cap I. \tag{2.39}$$

Therefore, regardless of the extent of the dilations, the reconstruction remains constrained by the mask, I. Typical unit-sized structuring elements are either a 4-connected or 8-connected neighborhood for a 2D image, and a 6-connected or 26-connected neighborhood for a 3D volume.

Formally, the reconstruction ρ_I of I from $J \subseteq I$ can be written as

$$\rho_I(J) = \bigcup_{n \geq 1} \delta_I^{(n)}(J), \tag{2.40}$$

where $\delta_I^{(n)}(J)$ represents n iterations of the elementary geodesic dilation operation, represented as

$$\delta_I^{(n)}(J) = \underbrace{\delta_I^{(1)} \diamond \delta_I^{(1)} \diamond \cdots \diamond \delta_I^{(1)}(J)}_{n \text{ times}}. \tag{2.41}$$

In this context, $\diamond$ denotes the use of the result from the previous step in subsequent iterations. For example, $\delta_I^{(1)} \diamond \delta_I^{(1)}(J)$ means that the initial geodesic dilation is performed, and the result of this step is then used in the next iteration to perform another iteration of the geodesic dilation operation.

For gray-scale images, the application of reconstruction is similar to the procedure described for binary images. If I and J are two images on the same discrete domain D, where each element (pixel or voxel) of the images is a value in the discrete set $\{0, 1, \cdots, L - 1\}$, such that $J(p) \leq I(p)$ for every pixel in the image, $p \in D$, the result of reconstruction is obtained by iterating the elementary geodesic dilation operation (in gray scale) until the stopping criterion is reached.

The elementary gray-scale geodesic dilation operation, under the condition that $J \subseteq I$, is defined as

$$\delta_I^{(1)}(J) = (J \oplus B) \wedge I, \tag{2.42}$$

where $\wedge$ represents the infimum, $J \oplus B$ is dilation, and B refers to the unit-sized structuring element. The reconstruction of image I from J, $\rho_I(J)$, can formally be defined as

$$\rho_I(J) = \bigvee_{n \geq 1} \delta_I^{(n)}(J), \tag{2.43}$$

where $\bigvee$ refers to the supremum.

Reconstruction may be deployed to achieve segmentation using fuzzy connectivity and used to perform extraction of different organs in CT images, as illustrated in Chapters 4, 6, and 7.

2.10.4 SEGMENTATION USING OPENING-BY-RECONSTRUCTION

According to Bloch [95], there exists an equivalence between the concept of connectedness as described in Section 2.9.3 and that of the degree of connectedness as defined by Udupa and Samarasekera [85]. As a result, the properties, transformations, and applications related to and derived from these two notions are similar [95]. Morphological reconstruction operates on the notion of

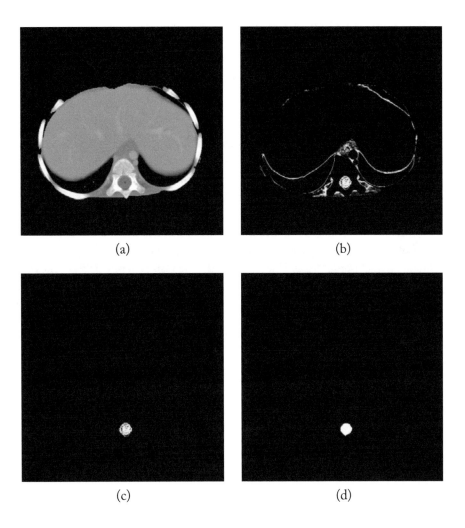

(a)

(b)

(c)

(d)

Figure 2.18: (a) A 512 × 512-pixel CT image to demonstrate the application of fuzzy connectivity using reconstruction. The slice is shown after contrast adjustment for improved viewing. The seed pixel to perform segmentation of the spinal canal is shown with an asterisk ("*"). (b) Fuzzy mapping of the image in (a) to identify the spinal canal. The parameters for the spinal canal were estimated as $\mu_s = +30$ HU and $\sigma_s = 17$ HU for use in Equation 2.23. (c) Reconstruction of the image in (a) with the marker seed. (d) Convex hull of the result after thresholding the reconstructed image in (c) at the membership value of 0.8.

a connection cost, or the minimum distance between specific points in a defined set. As a result, because of the similarity, reconstruction can be implemented in place of fuzzy connectivity.

An example of the application of reconstruction to achieve segmentation using fuzzy connectivity is shown in Figure 2.18. Here, reconstruction is used along with fuzzy mapping to perform segmentation of the spinal canal. The result of reconstruction was thresholded at a high membership value and closed using the convex hull to obtain the result shown in part (d) of Figure 2.18.

2.11   REMARKS

The segmentation of an image requires the delineation of homogeneous as well as inhomogeneous regions. Although many algorithms and techniques have been developed for image segmentation, it still remains a difficult and complex problem. Segmentation of medical images is particularly a difficult task, because the structures of interest vary greatly in terms of size and shape, and their HU values are affected by neighboring tissues as well as the inherent noise in the image. *Prior knowledge* about the characteristics of the organs of interest can assist in obtaining better results.

Several methods for image segmentation have been described in this chapter. The approaches presented have shown high potential, and have provided good results in practical applications. Methods of image segmentation and analysis, such as morphological image processing operators and deformable models, have been discussed in this chapter, and are employed in the procedures described in the following chapters.

CHAPTER 3

Experimental Design and Database

3.1 EXPERIMENTAL DESIGN

The CT exams used in this work, containing varying numbers of slices, are anonymous cases from the Alberta Children's Hospital. Two datasets were used in the work. The first dataset contains 10 CT images of four patients and the second dataset contains 30 CT images of 10 patients with neuroblastoma. The first dataset was used to perform quantitative evaluation of the result of identification and segmentation of different landmarks, such as the spinal canal and the diaphragm, and also to evaluate the effects of delineation of different organs and tissues in the segmentation of primary neuroblastic tumors. The second dataset was used to test all of the segmentation methods applied on the first dataset and the results were evaluated quantitatively with several exams in the case of segmentation of the vertebral column and the pelvic surface. The results of segmentation of neuroblastic tumors were evaluated qualitatively with the second dataset.

Both datasets contain images acquired with the *one-breath technique* and *two-breath technique*. In the one-breath technique, the patient is scanned from the top of the chest to the bottom of the pelvis in one breath hold. All slices are contiguous with no overlap. On the other hand, in the case of the two-breath technique, the patient is first scanned from the top of the chest to the bottom of the diaphragm in one breath hold with contiguous slices. Then, the table is repositioned at the top of the diaphragm and the patient is scanned to the bottom of the pelvis in a second breath hold with contiguous slices. As a result, there is overlap between the top and the bottom of the diaphragm between the two acquisitions, with both acquisitions combined into one series. For the present work, in the case of the two-breath technique, the images acquired in the first breath hold were discarded.

It is often difficult to perceive the spatial location and extent of 3D objects when they are viewed on a slice-by-slice basis. Therefore, it is beneficial to view CT datasets in 3D. To view and evaluate the results of the segmentation algorithms in 3D, a volume rendering software, *Advance Visual Systems/Express*® (AVS) [96], was used. For the purpose of 2D viewing and inspection, another visualization software, *ImageJ* [97], was used along with MATLAB. The computer used to process the exams is a Dell Precision PWS490 with an *Intel*® *Xeon*™ 3.00 GHz processor, 4 MB of cache memory, and 4 GB of RAM.

Approval was obtained from the Conjoint Health Research Ethics Board, Office of Medical Bioethics, University of Calgary, and the Calgary Health Region.

3.2   CT EXAMS AND DATASET

In the first dataset, the exams are of four patients of age two weeks to 11 years, with neuroblastoma at different stages of the disease and treatment, varying from a strongly heterogeneous tumor to a predominantly calcified tumor. Details of the patients, the CT exams, and brief descriptions of the tumors are summarized in Table 3.1.

In the present work, the term "diffuse" refers to a tumor mass with ill-defined boundaries. Although such a tumor may also be heterogeneous in tissue content, the primary distinguishing feature is the lack of clear boundaries. Heterogeneity refers to a tumor mass that is composed of all three primary tissue types; that is, there are significant quantities of viable tumor, necrotic tissue, and calcified tissue within the tumor mass.

The CT exams in the first dataset were acquired using a GE Medical System Lightspeed QX/i or a QX/i Plus helical CT scanner. All CT exams belonging to this dataset, except Exam 4C, include contrast enhancement, and have been used in the present work. The CT exams in the first dataset have an inter-slice resolution of 5 mm, and the intra-slice resolution varies from 0.35 mm to 0.55 mm.

Table 3.1: Description of the patients and the CT exams in the first dataset used in the present work. **Date** refers to when the CT exam was performed. All CT exams included the use of a contrast agent unless specified otherwise.

Patient	Exam	Age	Date	Tumor description
1, male	A	2 years	04/23/2001	large, heterogeneous
	B	2 years	06/07/2001	large, heterogeneous
	C	2 years	09/06/2001	small, calcified, homogeneous
2, female	A	2 years	03/31/2000	large, viable tumor & necrosis, diffuse
	B	2 years	07/21/2000	small, viable tumor & necrosis, diffuse
3, male	A	2 weeks	11/04/1999	small, viable tumor, diffuse
	B	3 months	02/08/2000	barely discernible, viable tumor, diffuse
4, female	A	10 years	02/22/2001	large, viable tumor, diffuse
	B	10 years	04/12/2001	small, viable tumor, diffuse
	C	10 years	05/17/2001	small, viable tumor & necrosis, diffuse; no contrast agent
	D	11 years	06/25/2001	small, viable tumor & necrosis, diffuse

In the second dataset, the exams are of 10 patients of age 14 months to 20 years, with neuroblastoma at different stages of the disease and treatment, varying from a strongly heterogeneous tumor to a predominantly calcified tumor. Details of the patients and CT exams are summarized in Table 3.2.

The CT exams in the second dataset were acquired using a GE Medical System Lightspeed Plus or VCT helical CT scanner. The CT exams have varying inter-slice resolution of 2.5 mm or 5 mm, and the intra-slice resolution varies from 0.39 mm to 0.70 mm. The CT exams of two patients are exceptional in terms of location of the primary tumor: the CT exam of Patient-8 contains a pelvic tumor, and the CT exams of Patient-11 contain thoracic tumors. All other patients have abdominal tumors. Some of the exams do not include contrast enhancement or any tumor at all (see Table 3.2). The age of Patient-9 is not typical, because neuroblastoma is a pediatric disease.

All of the CT exams were tested for automatic identification and segmentation of different landmarks. Using the landmarks, some organs were delineated to improve the definition of the tumor mass by the segmentation procedures described in Chapter 2.

3.3   METHODS OF EVALUATION OF THE RESULTS

Visual evaluation of the results of image processing is a highly subjective process. Although manual segmentation of different organs and tumors by experts provides the "gold standard" or "ground truth" for objective and quantitative assessment of the results of image processing, the approach is tedious, time-consuming, and subject to inter- and intra-operator variability. The accuracy achievable is limited mainly by the ability to define a precise boundary in the CT images. In order to overcome this difficulty and to aid the radiologist in evaluation and decision making, techniques for CAD have been developed.

The results of segmentation methods in this work were evaluated against manual detection and segmentation. Boundaries of organs of interest, ROIs, and tumors were drawn on several images by an expert radiologist (Dr. G. S. Boag). Most of the results were assessed by performing *qualitative* and *quantitative* evaluation. In some cases, only qualitative evaluation was performed as manual segmentation of the corresponding regions or organs is not practically feasible.

3.3.1   QUALITATIVE ASSESSMENT

The results of the identification and segmentation processes for different landmarks were assessed qualitatively by comparing them with standard anatomical atlases. Prior knowledge of anatomy was incorporated in visual inspection and analysis.

Table 3.2: Description of the patients and the CT exams in the second dataset used in the present work. **Date** refers to when the CT exam was performed. All CT exams included the use of a contrast agent, unless specified otherwise. The inter-slice resolution is 5 mm, unless specified otherwise.

Patient	Exam	Age	Date	Comments
5, male	A	3 years	03/30/2005	no contrast agent
	B	3 years	03/30/2005	
	C	3 years	04/21/2005	
	D	3 years	05/17/2005	
	E	3 years	08/02/2005	
6, male	A	2 years	03/16/2005	
	B	2 years	05/06/2005	
	C	2 years	07/12/2005	
7, female	A	3 years	11/21/2005	no contrast agent
	B	3 years	11/21/2005	
	C	3 years	12/15/2005	
	D	3 years	01/09/2006	
	E	3 years	04/12/2006	
8, female	A	14 months	05/31/2005	pelvic tumor, inter-slice resolution 2.5 mm
9, female	A	20 years	11/06/2006	exceptional in terms of age
	B	20 years	12/14/2006	no tumor, inter-slice resolution 2.5 mm
10, male	A	2 years	09/06/2006	
	B	2 years	10/23/2006	inter-slice resolution 2.5 mm
	C	2 years	12/21/2006	inter-slice resolution 2.5 mm
11, female	A	6 years	05/03/2005	chest scan only
	B	6 years	05/10/2005	scan of the thoracic tumor only
12, female	A	18 months	08/17/2006	no contrast agent
	B	18 months	08/17/2006	inter-slice resolution 2.5 mm
	C	19 months	10/04/2006	inter-slice resolution 2.5 mm
	D	21 months	12/12/2006	inter-slice resolution 2.5 mm
13, male	A	10 years	03/08/2006	
	B	10 years	04/25/2006	
	C	10 years	06/21/2006	
14, female	A	4 years	03/22/2007	no contrast agent, inter-slice resolution 2.5 mm
	B	4 years	03/22/2007	inter-slice resolution 2.5 mm

3.3.2   QUANTITATIVE ASSESSMENT

The results of segmentation of several organs or structures, such as the vertebral column, the spinal canal, the diaphragm, and the pelvic surface, have been assessed quantitatively in comparison with manual segmentation performed by an expert radiologist (Dr. G. S. Boag) using the Hausdorff distance [98] and the mean distance to the closest point (MDCP) [99]. The results of tumor segmentation were evaluated by comparing the computed volumes of segmentation of the primary tumor masses with those of the masses segmented manually by the radiologist. The total error rate as well as the false-positive error rate and true-positive rate of the results of segmentation (in percentage with respect to the corresponding manually segmented volume) were used to evaluate the accuracy of the procedures.

3.3.2.1   The Hausdorff Distance

The Hausdorff distance between two sets is the maximum of the distances of the points in one set to the corresponding nearest points in the other set [98]. More formally, the directed Hausdorff distance from set A to set B is a max-min function, defined as

$$h\left(A, B\right) = \max_{a \in A} \left[\min_{b \in B} \{d\left(a, b\right)\} \right],$$ (3.1)

where a and b are points in the sets A and B, respectively, and $d(a, b)$ is any metric between the points; for simplicity, $d(a, b)$ can be regarded as the Euclidean distance between a and b. If, for instance, A and B are two sets of points, a brute-force algorithm to calculate the Hausdorff distance is as follows:

1. $h = 0$;

2. for every point a_i of A,

   (a) *shortest distance* $= \infty$;

   (b) for every point b_j of B, $d_{ij} = d(a_i, b_j)$;
       if $d_{ij} <$ *shortest distance* then
       shortest distance $= d_{ij}$;

   (c) if *shortest distance* $> h$ then
       $h =$ *shortest distance*.

It should be noted that the Hausdorff distance is oriented (or asymmetric), which means that, in general, $h(A, B)$ is not equal to $h(B, A)$.

A more general definition of the Hausdorff distance is

$$H\left(A, B\right) = \max\{h\left(A, B\right), h\left(B, A\right)\},$$ (3.2)

which defines the Hausdorff distance between A and B. In comparison, Equation 3.1 gives the Hausdorff distance from A to B (also called the directed Hausdorff distance). The two distances

$h(A, B)$ and $h(B, A)$ are, sometimes, termed as the forward and the backward Hausdorff distances of A to B. Although the terminology is not consistent among authors, Equation 3.2 is commonly used to indicate the Hausdorff distance.

The Hausdorff distance measures the degree of mismatch between two sets by calculating the distance of the point in A that is the farthest from any point in B, or vice versa. Intuitively, if the Hausdorff distance is h, then every point in A must be within the distance h of some point in B, and vice versa. The Hausdorff distance is a popular measure to compare the results of segmentation of ROIs in medical images [100]. In this work, the Hausdorff distance is used to measure the accuracy of the results of segmentation of the vertebral column, the spinal canal, the diaphragm, and the pelvic girdle.

3.3.2.2 Mean Distance to the Closest Point (MDCP)

The Hausdorff distance measures the mismatch between two sets that are at fixed positions with respect to one another, and represents the worst case in matching the two sets. Another measure of comparison called the *Mean Distance to the Closest Point* (MDCP) has been proposed to provide a more general measure than the Hausdorff distance [99]. MDCP evaluates quantitatively the degree of affinity between two sets by computing the average distance to the closest point (DCP) between the two sets (see Saha et al. [101] for a related measure).

Given two sets, $A = \{a_1, a_2, \ldots, a_M\}$ and $B = \{b_1, b_2, \ldots, b_N\}$, DCP is defined as

$$\text{DCP}\,(a_i, B) = \min \|a_i - b_j\|, \quad j = 1, 2, \ldots, N, \tag{3.3}$$

where $\|.\|$ is some norm (for example, the Euclidean norm) computed for the points a_i and b_j. MDCP is then defined as

$$\text{MDCP}\,(A, B) = \frac{1}{M} \sum_{i=1}^{M} \text{DCP}\,(a_i, B). \tag{3.4}$$

The smaller the value of the MDCP is, the more similar the two sets are to each other. It should be noted that, although MDCP is a useful measure, it is not a metric.

In the present work, MDCP is used to compare the contours of the vertebral column, the spinal canal, the diaphragm, and the pelvic girdle obtained by the segmentation procedures for each slice, with the corresponding contour drawn independently by the radiologist.

3.3.2.3 Measures of Volumetric Accuracy of Tumor Masses

To evaluate the results of tumor segmentation and the benefits of incorporating the surface of the diaphragm or of the pelvis in the procedures for segmentation of the neuroblastic tumor, the results were compared against the primary tumor volume definition, as segmented by an expert pediatric radiologist, using the measures of *total error rate* (ε_T), *false-positive error rate* (ε_{FP}), and *true-positive rate* (ε_{TP}).

The total error rate (ε_T) is defined as

$$\varepsilon_T = \frac{V(A) - V(R)}{V(R)} \times 100\%, \tag{3.5}$$

where $V()$ is the volume, A is the result of segmentation using the segmentation procedures, and R is the result of segmentation by the radiologist (the ground truth).

A false positive (FP) is a voxel that has been included in the final result of segmentation, but is not actually a part of the tumor mass as defined by the radiologist. The false-positive error rate (ε_{FP}) is defined as

$$\varepsilon_{FP} = \frac{V(A) - (V(A) \cap V(R))}{V(R)} \times 100\%. \tag{3.6}$$

A true positive (TP) is a voxel that has been included in the final result of segmentation, and is actually a part of the tumor as defined by the radiologist. This metric is an indication of the effectiveness of the procedure in capturing true tumor voxels in the final result of segmentation. The true-positive rate (ε_{TP}) is defined as

$$\varepsilon_{TP} = \frac{V(A) \cap V(R)}{V(R)} \times 100\%. \tag{3.7}$$

3.4   REMARKS

Evaluation of the results of segmentation of medical images is an important but difficult task. Without the application of several types of evaluation procedures, it could be impossible to assess the accuracy and the effectiveness of the segmentation processes. In the practical world, it is difficult to compare or evaluate the results by involving many expert radiologists, and to use all available measures or metrics to examine the accuracy of the results in all aspects.

In this chapter, a few methods for evaluation and comparison of results of segmentation were described. The datasets and the experimental set up used in related studies were also provided. Several image processing and segmentation methods have been deployed to perform landmarking and segmentation of different organs with the datasets, and the results have been assessed using the evaluation processes described in this chapter. The following chapters provide examples of the results obtained in related studies.

CHAPTER 4

Segmentation of the Ribs, the Vertebral Column, and the Spinal Canal

4.1 THE VERTEBRAL COLUMN AND THE SPINAL CANAL

The vertebral column plays a central role in the human biomechanical system [38, 43]. Many factors, such as herniated discs, degenerative diseases like osteoporosis, vertebral neoplasms, accidental injuries, and scoliosis can derange or restrict this vulnerable system [38, 43]. In diagnosis, therapy, and surgical intervention, segmentation of the vertebral column in CT images is a crucial preprocessing step. Due to the complexity of the vertebral surface, small distances between the articulated vertebrae, and above all, the presence of pathological regions such as tumors and calcification, segmentation of the vertebral column and the ribs are challenging tasks.

Accurate planning of radiation therapy needs the definition of treatment volumes and a clear delimitation of normal tissue of which exposure should be prevented or minimized [28, 41, 102]. The spinal canal is a radio-sensitive organ, and it should be precisely identified to prevent complications arising from radiation-induced damage. The automatic detection of the spinal cord in CT images is a difficult problem, and while there has been much work done on registration and segmentation of images of the spine [28, 38, 41, 43], the development of easily applicable, robust, and automatic registration and segmentation algorithms remains difficult.

Various algorithms have been developed to identify the spine or the vertebral column [38, 43, 61], the spinal canal [28, 41, 42, 61], and the ribs [39, 40]. In this context, methods based on morphological image processing are described in the present chapter to perform automatic segmentation of the ribs, the vertebral column, and the spinal canal in pediatric CT exams.

4.2 REMOVAL OF PERIPHERAL ARTIFACTS AND TISSUES

In the segmentation procedures presented in this work, each CT volume is first processed to identify and remove peripheral artifacts, the skin, the peripheral fat, and the peripheral muscle using the procedures described by Deglint et al. [11], Rangayyan et al. [42], Vu [103], and Vu et al. [12]. The procedures are discussed in the following sections in brief.

4.2.1   REMOVAL OF THE EXTERNAL AIR

Air, by definition, has a CT number of -1000 HU. To remove the air external to the body, the CT volume is thresholded with the range -1200 HU to -400 HU to account for variations due to noise and partial-volume averaging. 2D binary reconstruction using a 4-connected neighborhood is applied to each slice of the CT volume, where the four corners of each slice are used as the markers and each thresholded slice is the mask. The procedure is applied on a slice-by-slice basis because the results of a 3D approach were observed to leak into the bowels, in some cases, due to the low spatial resolution of the image data. Once reconstruction is completed, the resulting volume is morphologically closed using a disk-shaped structuring element of radius 10 pixels (approximately 5 mm) to remove material external to the body, such as the patient table, blanket, and tubes connected to intravenous drips.

4.2.2   REMOVAL OF THE SKIN

The first expected layer from the outside of the body is the skin, which usually has a thickness of $1 - 3$ mm [42]. Using the parameter of the expected skin thickness, the boundary of the body previously obtained via segmentation of the air region could be shrunk using a 3D morphological operator to include the skin. Instead, in the present work, the air region is dilated in 2D using a disk-shaped structuring element of radius 2 pixels to separate the skin layer from the body. The skin boundary can be used as a landmark for registration and segmentation of medical images [104].

4.2.3   REMOVAL OF THE PERIPHERAL FAT

The next layer after the skin is the peripheral fat. Fat has a mean CT value of $\mu = -90$ HU with $\sigma = 18$ HU [105, 18]. Peripheral fat around the abdomen varies in thickness from 3 mm to 8 mm in children. After removal of the skin, voxels within a distance of 8 mm from the inner skin boundary are examined for consideration as fat. If these voxels fall within the range of $-90 \pm 2 \times 18$ HU, they are classified as peripheral fat.

4.2.4   REMOVAL OF THE PERIPHERAL MUSCLE

The anatomical structures of skin and peripheral fat were not observed to interfere in the segmentation of the neuroblastic tumor; however, their segmentation and removal provides the easiest route to the peripheral muscle, a structure that was observed to get mixed with the segmented tumor volume [11, 12]. Peripheral muscle has a mean CT value of $\mu = +44$ HU with standard deviation $\sigma = 14$ HU [18, 105], and a thickness ranging from 6 mm to 10 mm in abdominal sections. Similar to the procedure for the removal of the peripheral fat, voxels found within 10 mm of the inner fat boundary and within the range of $44 \pm 2 \times 14$ HU are classified as peripheral muscle. Then, the peripheral fat region obtained is dilated using a disk-shaped structuring element of radius 2 mm to remove the discontinuities and holes between the peripheral fat region and the peripheral muscle, and are included as parts of the peripheral muscle.

The procedures for delineating the peripheral artifacts and tissues are summarized in the flowchart shown in Figure 4.1. Results demonstrating the growth of the air region, peripheral fat, and peripheral muscle are shown in Figure 4.2. The peripheral structures have been removed by setting the corresponding elements in the original image to an arbitrarily low value of −3024 HU. Note that, in Figure 4.2 (f), the peripheral muscle has been accurately removed, leaving the bone structure and the other organs intact. 3D illustrations of the surfaces obtained at each stage are shown in Figure 4.3.

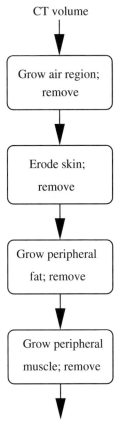

Figure 4.1: Flowchart describing the preprocessing steps.

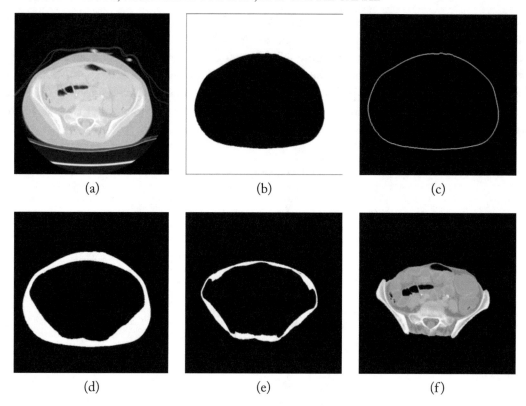

(a)   (b)   (c)

(d)   (e)   (f)

Figure 4.2: (a) A 512×512-pixel cross-sectional CT slice from Exam 4D, with the display range $[-1050, 400]$ HU. Note that the patient's table and blanket are visible outside the body. (b) External air and artifacts are shown in white. The remaining part of the image is shown in black. (c) The skin layer. (d) The peripheral fat region. (e) The peripheral muscle region. (f) The result after removing the external air and artifacts, the skin, the peripheral fat, and the peripheral muscle. Reproduced with kind permission of Springer Science+Business media and from R M Rangayyan, S Banik, and G S Boag. "Landmarking and segmentation of computed tomographic images of pediatric patients with neuroblastoma." *International Journal of Computer Assisted Radiology and Surgery*, 2009. In press. © Springer. See Figure 4.3 for related 3D illustrations.

4.3   IDENTIFICATION OF THE RIB STRUCTURE

The segmentation procedure to detect and delineate the ribs in pediatric CT images is initialized by removing the external air, peripheral artifacts, and the skin layer, and thresholding each CT volume at 200 HU. Then, the binarized volume is morphologically opened using a disk of radius 3 pixels to disconnect the rib structure from the spine. Because the ribs in the upper thoracic region are smaller and closer to one another than in the middle of the thorax, the compactness factor, defined

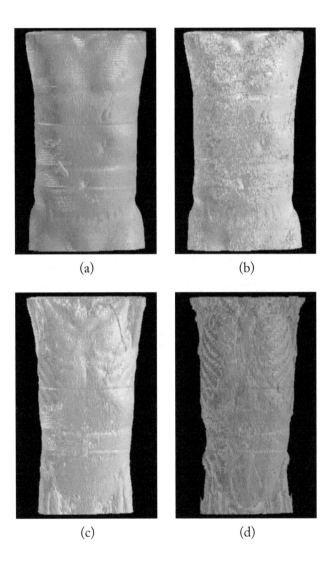

(a)   (b)

(c)   (d)

Figure 4.3:   Illustrations to demonstrate the removal of the external air, peripheral artifacts, the skin layer, the peripheral fat, and the peripheral muscle in Exam 4D. 3D representation of the body: (a) after removal of the external air and peripheral artifacts; (b) after removal of the skin layer; (c) after removal of the peripheral fat region; and (d) after removal of the peripheral muscle. Reproduced with kind permission of Springer Science+Business media and from R M Rangayyan, S Banik, and G S Boag. "Landmarking and segmentation of computed tomographic images of pediatric patients with neuroblastoma." *International Journal of Computer Assisted Radiology and Surgery*, 2009. In press. © Springer.

as $cf = 1 - 4\pi A/P^2$, where A is the area and P is the perimeter of the corresponding region, of the inner contour of the peripheral fat region on each slice is used to differentiate between the upper portion of the thoracic region and the rest of the thorax as well as the abdomen. Because cf is a measure of the complexity of the shape of a region, starting from the top-most slice of a CT scan, the first slice with $cf < 0.4$ is considered to be the end of the upper portion of the thoracic region present in the CT volume.

For each slice, a central line along the medial sagittal plane is defined based on the inner contour of the peripheral fat region. The use of the information related to the peripheral fat boundary aids in defining the central line close to the mid-sagittal plane regardless of the position of the body in the image. An example of the defined central line is shown in red in Figure 4.4 (a). Note that the defined central line passes through the spine, as expected.

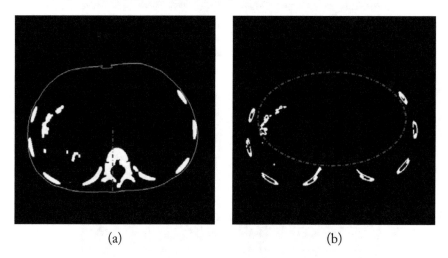

(a)   (b)

Figure 4.4: (a) A 512×512-pixel CT slice after thresholding at 200 HU. The defined central line is shown in red. The inner edge of the peripheral fat region is shown for reference. (b) After the application of reconstruction procedure, the boundary of the defined elliptical region to remove unwanted structures inside the body is shown in red. Calcified parts of a tumor due to neuroblastoma are visible, and are removed from the result of segmentation of the rib structure by this procedure. Reproduced with kind permission of Springer Science+Business media and from S Banik, R M Rangayyan, and G S Boag. "Automatic segmentation of the ribs, the vertebral column, and the spinal canal in pediatric computed tomographic images." *Journal of Digital Imaging*, 2009. In press. © SIIM.

Then, an initial estimation procedure is applied as follows [60]:

1. Consider each slice starting from the top of the binarized CT volume and mark each of the thresholded regions on each slice.

2. Define the central line.

3. Measure the shortest Euclidean distance (ED) from the edge and the centroid of each of the regions to the defined central line and also to the inner edge of the peripheral fat region.

4. If the corresponding slice is not from the upper portion of the thoracic region (as defined previously using the measure of cf), consider a marked region initially to be a part of the rib structure if the region satisfies all of the following criteria:

    (a) Shortest ED from the centroid of the region to the inner edge of the peripheral fat region ≤ 3.5 cm.

    (b) Shortest ED from the edge of the region to the inner edge of the peripheral fat region ≤ 1.8 cm.

    (c) Shortest ED from the centroid of the region to the defined central line ≥ 1.5 cm.

    (d) Shortest ED from the edge of the region to the defined central line ≥ 1.5 cm.

5. If the corresponding slice is from the upper portion of the thoracic region, consider a marked region initially to be a part of the rib structure if the corresponding region satisfies all of the following criteria:

    (a) Shortest ED from the centroid of the region to the inner edge of the peripheral fat region ≤ 4.5 cm.

    (b) Shortest ED from the edge of the region to the inner edge of the peripheral fat region ≤ 3 cm.

    (c) Shortest ED from the centroid of the region to the defined central line ≥ 1 cm.

    (d) Shortest ED from the edge of the region to the defined central line ≥ 1 cm.

6. Calculate the area, the minor axis length (a measure of thickness of the corresponding region on a slice), and cf of each of the selected regions in the corresponding slice. If, for a given region, $cf \leq 0.85$, area of the region ≤ 5000 pixels, and the minor axis length ≤ 2 cm, accept the region to be a part of the rib structure.

7. Continue the process until the lowest slice that contains possible parts of the rib structure is processed.

The above-mentioned spatial distance limits were obtained through experimentation with the full dataset, as described in Section 3.2. The structure obtained with the above-mentioned procedure can be considered as a primary estimation of the rib structure; the results could include some unwanted structures that do not belong to the rib structure.

The initially detected rib structure is skeletonized, and the skeleton of the rib structure is used as a region marker to perform reconstruction in 3D using the 26-connected neighborhood; the result is thresholded at a high fuzzy membership value of 0.8. The result of reconstruction may still include some other calcified structures in the abdomen or in the thorax.

To eliminate unwanted structures inside the body, a 2D elliptical region is defined on each slice inside the thoracic and abdominal regions based on the outer boundary of the initially detected rib structure and the inner contour of the peripheral fat region for the corresponding slice. The center, the major axis, and the minor axis are chosen in such a manner that the elliptical region fits inside the initially segmented rib structure for each slice. To obtain the center of the defined elliptical region, the center point of the inner contour of the peripheral fat region is shifted toward the anterior of the body by 10% of the maximum distance along the sagittal plane between the anterior and posterior side of the inner contour of the peripheral fat region on each slice, and is considered to be the center of the defined elliptical region for the corresponding slice. The shift is required to ensure that parts of the rib structure close to the spine are not removed. The major axis length is defined to be 90% of the maximum distance between the ribs on the left and the right side (along the coronal plane); the minor axis length is defined to be 80% of the maximum distance along the sagittal plane between the anterior and posterior side of the inner contour of the peripheral fat region. All pixels inside the defined elliptical region are removed. An example of the defined elliptical region is shown in Figure 4.4 (b). This procedure also helps in finding the lower limit of the rib structure in the CT exam, by eliminating the abdominal calcified regions which could create ambiguity in the process of detection of the rib structure.

After removing the unwanted structures inside the thoracic and abdominal regions, the features used in the initial estimation step are applied again to differentiate between the ribs and other regions. Finally, after removing the voxels corresponding to other regions, and filling in holes, the resulting volume is dilated in 2D using a disk-type structuring element of radius 2 pixels, to obtain a refined segmentation of the rib structure.

4.3.1 ASSESSMENT OF THE RESULTS

The procedure for the detection of the rib structure was tested with 39 CT exams of 13 patients; Exam 8A was not considered for the rib segmentation process because it contains only a scan of the pelvic region. The segmented rib structure obtained by the procedure described was observed to be a satisfactory representation of the actual rib structure. Two examples of the segmented ribs from Exams 4D and 1C, in 2D, are shown in Figure 4.5; the 3D representation of the segmented rib structure for two CT exams (Exams 4D and 14B) are shown in Figure 4.6.

The procedure produced good results irrespective of the position of the patient in the image, and did not include any part of the spine in any of the 39 CT exams processed. The process did not include any calcified tumor region or any other ambiguous region close to the rib structure, except in two CT exams, in which calcified portions of the tumor were spatially connected to the ribs.

Because the regions inside the defined elliptical boundary were removed from consideration, the methods detected precisely the lower limit of the rib structure by eliminating the possibilities of including other regions in the result of segmentation in all CT exams processed. At the top of thoracic region, the results included the scapula (shoulder blade), the clavicle (collar bone), and the top of the humerus (arm bone), in almost all the cases, if they were present in the corresponding

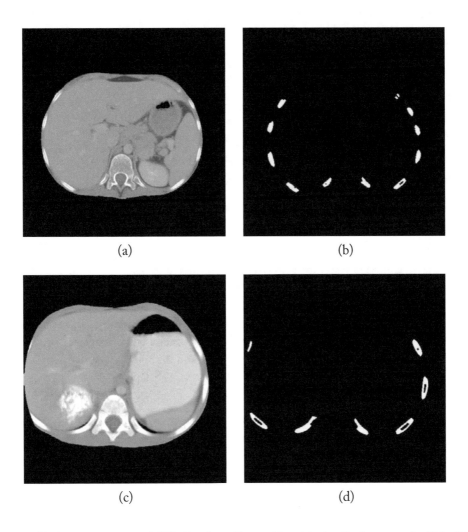

(a)   (b)

(c)   (d)

Figure 4.5: (a) A 512 × 512-pixel CT slice from Exam 4D, with the display range [−200, 400] HU. (b) Detected ribs in the slice shown in part (a). See Figure 4.6 (a) for a related 3D illustration. (c) A 512 × 512-pixel CT slice from Exam 1C; display range [−200, 400] HU. (d) Detected ribs in the slice shown in part (c). Note that the procedure has detected the rib structure satisfactorily, regardless of the patient's position, and with the presence of a calcified tumor close to the ribs. [See the lower left-hand side of the image in part (c)]. Reproduced with kind permission of Springer Science+Business media and from S Banik, R M Rangayyan, and G S Boag. "Automatic segmentation of the ribs, the vertebral column, and the spinal canal in pediatric computed tomographic images." *Journal of Digital Imaging*, 2009. In press. © SIIM.

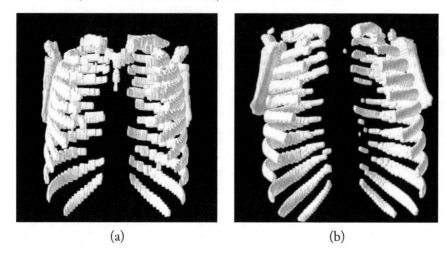

(a)   (b)

Figure 4.6: Two examples of segmented rib structures, displayed in 3D. (a) Detected rib structure for Exam 4D. See Figure 4.5 (a) for related illustration of a selected slice. (b) Detected rib structure for Exam 14B. Note that some parts of the ribs that are close to the vertebral column have not been included in the result. The scapula and the clavicle are included in the results of segmentation. Reproduced with kind permission of Springer Science+Business media and from S Banik, R M Rangayyan, and G S Boag. "Automatic segmentation of the ribs, the vertebral column, and the spinal canal in pediatric computed tomographic images." *Journal of Digital Imaging*, 2009. In press. © SIIM.

CT exam. The procedure does not include steps to eliminate these structures from the result of segmentation because their removal is not relevant in the present work on landmarking. In some cases, where parts of the ribs were totally fused with the spine, they were not included in the results of segmentation (see Figure 4.6). In two CT exams, Exams 3A and 3B, the patient being of age two weeks and three months, respectively, some parts of the rib structure were not detected because of the small size of the patient, and the low spatial resolution of the image data.

Other methods for segmentation of the rib structure reported in the literature [39, 40, 41] have been able to produce good results. However, most of the published procedures are only applicable to CT scans of adults, where the ribs are well developed, and cannot be applied to pediatric CT exams. In pediatric cases, the ribs are not well developed, the bones are not fused together, and bony structures possess lower HU values than those of adults. In addition, the datasets used in the present work contain tumors and calcification in thoracic and upper abdominal regions that make the segmentation process more complex. Although some of the spatial distance limits used in the procedures were obtained through experimentation, the procedures described have produced good results with a wide variety of CT scans of patients of age varying from two weeks to 20 years.

4.4   SEGMENTATION OF THE VERTEBRAL COLUMN

For segmentation of the vertebral column, the CT volume is thresholded at 180 HU and the result is morphologically opened using a tubular structuring element of radius 2 pixels and height 3 pixels. The segmented rib structure obtained using the procedures described in Section 4.3 is eliminated from the binarized volume. Then, the remaining volume is morphologically eroded in 2D by a disk of radius 3 pixels to disconnect the connected components.

For each slice, a central line up to the mid-point of the body along the medial sagittal plane is defined based on the inner contour of the peripheral fat region; the defined central line is expected to pass through the vertebral column, as shown in Figure 4.7 (b). The relative position and the orientation of the body are taken into account by this step. For each slice, if any pixel in any region is within the Euclidean distance of 8 mm from the central line, the region is initially considered to be a part of the vertebral column. Then, the resulting image is evaluated in 3D and the longest bony structure along the inter-slice direction is considered to be the vertebral column.

Following the procedure as described above, the gradient of the detected binary volume was computed, binarized for all nonzero gradient magnitude values, and then added to the detected binary volume to create a combined mask of the vertebral column to reduce the possibilities of missing parts with low HU values [60]. The gradient helps to include the low-HU voxels in the vertebral column [see part (d) of Figure 4.7]. The image within the combined mask is then thresholded at 150 HU; to minimize the errors due to the partial-volume effect, the pixels within the combined mask are evaluated using a 5×5 window; the maximum and minimum are calculated, and if their difference is above 50 HU, then the center pixel is examined to find if it is within the range related to the vertebral column. If the constraints mentioned above are satisfied, the center pixel is included in the result of segmentation. Finally, the segmented vertebral column is morphologically closed using a disk of radius 2 pixels to remove isolated pixels and perform smoothing. Part (f) of Figure 4.7 shows the final result of segmentation for the CT slice shown in part (a) of the same figure. A 3D rendition of the vertebral column for the same CT exam is shown in Figure 4.8 (b).

4.4.1   QUALITATIVE EVALUATION OF THE RESULTS

The segmentation methods were applied to 40 CT images of 14 patients (see Section 3.2); Exam 8A was processed without using the information related to the rib structure because the exam contains only the pelvic region.

The results of segmentation were observed to be good representations of the vertebral column, excluding the pelvic girdle and the rib structure. Three examples of the 3D representation of the detected vertebral column are presented in Figure 4.8, along with the segmented spinal canal. The methods performed well by minimizing errors due to the partial-volume effect and including the parts of the vertebrae with low HU values. The strength of the procedure lies in segmenting the parts of the vertebral column which are barely identifiable in the corresponding CT slices because of low HU values due to poor inter-slice resolution and the partial-volume effect. The results did not include any tumoral or other tissues in thoracic and abdominal regions in any of the CT exams.

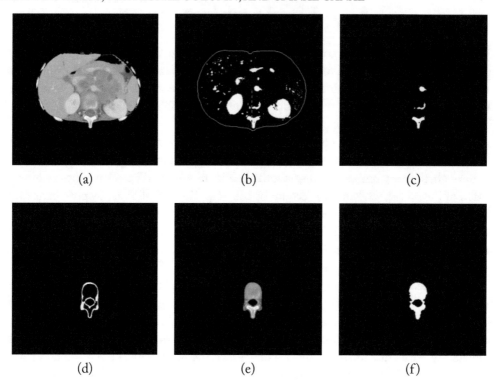

Figure 4.7: (a) A 512 × 512-pixel CT slice from Exam 4A, with the display range [−400, 500] HU. (b) The CT slice shown in part (a) after thresholding at 180 HU and removing the rib structure. The defined central line is shown in red. The inner edge of the peripheral fat region is shown for reference. (c) Initial result of segmentation. (d) The binarized gradient magnitude of the result of segmentation considering only the longest bony structure in the initial result. (e) The region separated after applying the combined mask. (f) Final result of segmentation. Reproduced with kind permission of Springer Science+Business media and from S Banik, R M Rangayyan, and G S Boag. "Automatic segmentation of the ribs, the vertebral column, and the spinal canal in pediatric computed tomographic images." *Journal of Digital Imaging*, 2009. In press. © SIIM. See Figure 4.8 (b) for a related 3D illustration.

Segmentation of the lower end of the vertebral column poses a great challenge. In six CT exams, the segmentation method failed to include precisely the small parts of the vertebrae in the lower spinal region, such as some portions of the sacrum or the coccyx; this was mainly due to the use of morphological erosion to eliminate the connected parts of the bowels with higher HU values than expected because of the use of the contrast agent, and the fusion of the pelvic girdle with the lower end of the spine. In five CT exams, the result included some parts of the bowels or the intestine,

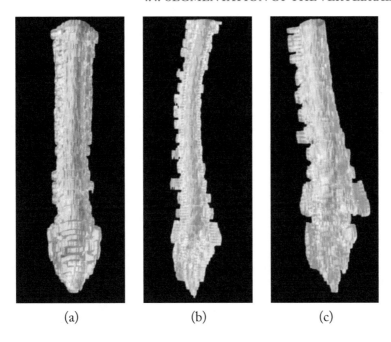

(a) (b) (c)

Figure 4.8: Three examples of the result of segmentation of the vertebral column and the spinal canal. The vertebral columns are shown in gray and the spinal canals are shown in red. (a) Exam 1A. (b) Exam 4A. (c) Exam 4B. Reproduced with kind permission of Springer Science+Business media and from S Banik, R M Rangayyan, and G S Boag. "Automatic segmentation of the ribs, the vertebral column, and the spinal canal in pediatric computed tomographic images." *Journal of Digital Imaging*, 2009. In press. © SIIM.

because of their inseparable appearance with the lower end of the spine in the corresponding CT images.

The segmentation method did not perform well in segmenting the vertebral column in two CT exams (Exam 9A and Exam 9B), by including some parts of the pelvic girdle; the patient is of 20 years of age, an exception in the dataset of pediatric CT exams in the study, and the pelvic girdle appeared to be merged with the spine. In five CT exams, in the thoracic spine region, minor leakage of the segmented region to the ribs resulted; in these cases, the thoracic vertebrae and the ribs cannot be separated even by the human eye [see part (c) of Figure 4.8].

4.4.2 QUANTITATIVE EVALUATION OF THE RESULTS

The Hausdorff distance and MDCP measures were used to compare the contours of the vertebrae obtained by the segmentation method for each CT slice, with the corresponding contour drawn independently by the radiologist, for 13 CT exams of six patients. The results are listed in Table 4.1.

The Hausdorff distance was computed for some of the selected CT slices where a section of the vertebral column was present, for each CT exam listed in Table 4.1. The minimum, the maximum, the average, and the SD were computed for each CT exam, over the selected 2D slices containing a section of the vertebral column. The mean Hausdorff distance for the corresponding CT exam was determined by taking the average of the Hausdorff distance over the selected individual 2D slices.

Table 4.1: Quantitative evaluation of the result of segmentation of the vertebral column. All distances are in millimeters. Mean = average, SD = standard deviation, min = minimum, and max = maximum. Reproduced with kind permission of Springer Science+Business media and from S Banik, R M Rangayyan, and G S Boag. "Automatic segmentation of the ribs, the vertebral column, and the spinal canal in pediatric computed tomographic images." *Journal of Digital Imaging*, 2009. In press. ©SIIM.

Exam	Number of slices	Hausdorff Distance				MDCP			
		min	max	mean	SD	min	max	mean	SD
5B	25	0.96	4.87	2.04	0.95	0.29	0.91	0.46	0.15
5C	26	1.38	15.16	3.97	3.69	0.32	3.03	0.75	0.67
5D	30	0.99	17.69	3.09	3.34	0.31	2.74	0.60	0.48
5E	27	1.10	9.01	3.62	2.20	0.32	7.63	1.10	1.76
6A	24	1.08	13.27	3.86	2.83	0.32	8.64	1.05	1.68
6B	28	1.09	5.85	3.06	1.51	0.33	1.03	0.60	0.21
6C	30	0.99	12.07	2.73	2.33	0.31	2.16	0.61	0.48
7B	32	0.95	4.86	2.02	0.90	0.29	0.71	0.50	0.10
7C	32	0.97	17.83	3.03	2.98	0.35	3.23	0.61	0.49
7D	36	0.96	12.22	2.92	2.49	0.29	3.51	0.73	0.66
12B	60	1.03	7.63	2.63	1.58	0.29	2.56	0.63	0.39
13A	49	1.19	25.18	5.24	4.89	0.36	4.76	1.09	1.01
14B	59	1.05	8.03	3.03	1.55	0.35	0.73	0.73	0.30

The MDCP was calculated on the same selected slices of the dataset, where a section of the vertebral column was present; the minimum, the maximum, the average, and the SD were calculated over all the selected slices for each CT exam. Over the 13 CT exams processed of the six patients listed in Table 4.1, a total of 458 CT slices were used to perform the quantitative evaluation.

The maximum Hausdorff distance and the maximum MDCP for each of the CT exams were calculated by considering the maximum of the corresponding measures over all of the selected 2D slices in the exam. The maximum values are significantly larger than the corresponding average measures in all CT exams because of isolated cases of slices in which there was an overestimation or an underestimation of the vertebrae. This is indicated by the large deviation in the results for a given exam, as shown in Table 4.1.

The 95% confidence interval, obtained by the one-sample two-tailed t-test [106], is [8.18, 15.46] for the maximum Hausdorff distances listed in the 4th column of Table 4.1. For the mean MDCP, listed in the 9th column of Table 4.1, the corresponding interval is [0.60, 0.86]. The average MDCP for the 13 exams processed, listed in Table 4.1, is 0.73 mm, and is of the order of the intra-slice resolution of the CT image. The average error (in terms of MDCP), in relation to the representative dimension of the vertebral column in the lumbar region (60 mm) for pediatric patients, is approximately 1.2%.

The errors in terms of the average Hausdorff distance and the average MDCP are small and indicate that the segmentation procedure is able to produce good results. In general, the results were found to be in good agreement with the corresponding result of manual segmentation. The contours obtained by the method described and the contours obtained by the manual segmentation overlapped completely in almost all cases. However, the error was observed to be high in the lower vertebral region because of the complex structure of the vertebral column in the pelvic region, poor spatial resolution, the partial-volume effect, the effect of the contrast agent, and the disjoint appearance of the vertebrae in the CT images.

The results of segmentation for four representative CT slices from different sections of the body are illustrated in Figure 4.9. The corresponding Hausdorff distance and MDCP are also provided. Parts (a), (b), and (c) in Figure 4.9 illustrate CT slices where the segmentation method performed well; the contours obtained by the method and the contours drawn by the radiologist overlap almost completely, producing small errors. In part (d) of the same figure, an example in which the segmentation method included parts of structures other than the vertebral column is displayed; the error is relatively high in this case.

4.5   IDENTIFICATION OF THE SPINAL CANAL

The procedure to detect the spinal canal proposed by Rangayyan et al. [42] is modified in the present work to make the process of automatic extraction of the spinal canal region robust and adaptive. The method operates on the observations that the spinal canal is homogeneous in CT characteristics, contrasts strongly with the surrounding bony structure of the vertebral column, is almost triangular in the lumbar and cervical regions, and is nearly circular in the thoracic regions [107]. Instead of the fuzzy connectivity algorithm used by Rangayyan et al. [42], reconstruction is deployed to grow the region within the vertebral foramen to perform segmentation of the spinal canal. In order to achieve automatic segmentation, the segmented vertebral column and the rib structure are taken as reference, and the Hough transform is used to detect seed voxels in the spinal canal. The process is described in the following sections.

4.5.1   DELIMITATION OF THE SEARCH RANGE FOR SEED DETECTION

An adaptive approach is followed for the purpose of detection of the seeds voxels for the spinal canal. The segmented vertebral column is used to delimit the search range for seed voxels. Because the spinal canal is contained within the vertebral column, a rectangular window is defined on each

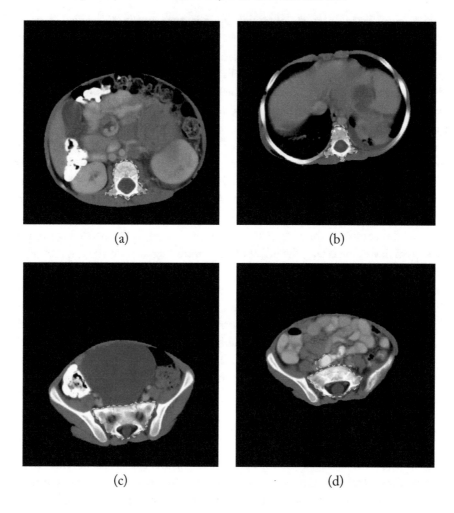

(a)   (b)

(c)   (d)

Figure 4.9: Four 512 × 512-pixel CT slices, illustrating qualitative and quantitative analysis of the results of segmentation. The solid red line denotes the contour of the vertebrae drawn independently by the radiologist, and the dashed green line represents the contour of the vertebrae produced by the segmentation method. (a) A slice from the abdominal region in Exam 5B. MDCP = 0.3 mm. Hausdorff distance = 1.4 mm. (b) A slice from the thoracic region in Exam 5C. MDCP = 0.4 mm. Hausdorff distance = 1.5 mm. (c) A slice from the pelvic region in Exam 5B. MDCP = 0.5 mm. Hausdorff distance = 2.4 mm. (d) A slice from the pelvic region in Exam 5C. MDCP = 0.7 mm. Hausdorff distance = 4.4 mm. Note that the result included a part outside the vertebral column. Figures in part (a), (b), and (c) reproduced with kind permission of Springer Science+Business media and from S Banik, R M Rangayyan, and G S Boag. "Automatic segmentation of the ribs, the vertebral column, and the spinal canal in pediatric computed tomographic images." *Journal of Digital Imaging*, 2009. In press. © SIIM.

slice in such a manner that the segmented vertebra in the corresponding slice completely fits inside the rectangular window. Because the spinal canal is nearly circular in the thoracic region [42, 107], to limit the scope of application of the Hough transform, only the thoracic region bounded by the highest and the lowest portions of the rib structure is considered for seed selection.

4.5.2   DETECTION OF SEED VOXELS USING THE HOUGH TRANSFORM

The Hough transform for the detection of circles, as given in Equation 2.20, is applied to the edge map of the cropped region containing the binarized vertebral column to detect the center of the best-fitting circle for each slice, with the radius limited to the range of 6 to 10 mm [42]. The edge map is obtained by applying Canny's procedure for edge detection, as described in Section 2.5. Because of the sparseness of the edge map, the global maximum in the Hough space may relate to the external curve of the anterior parts of the vertebra (that is, the vertebral body) instead of the desired circular boundary of the spinal canal inside the vertebra. To obtain the center and radius of the desired circle, the intensity values of bone marrow ($\mu = +142$ HU and $\sigma = 48$ HU) [42] and the spinal canal ($\mu = +23$ HU and $\sigma = 15$ HU) are used. The values of μ and σ for the spinal canal were manually estimated from 10 CT images. If the HU value of the voxel at the center of the best-fitting circle for a given slice is not within the range of $\mu \pm 2\sigma$ of the spinal canal (that is, within the range $[-7, 53]$ HU), the corresponding circle is rejected, and the next maximum in the Hough space is evaluated. This process is continued until a suitable center is determined. The process of seed detection is illustrated in Figures 4.10, 4.11, 4.12, and 4.13.

Figure 4.10 (d) shows the best-fitting circle drawn on a CT slice, shown in part (a) of the same figure. The related Hough space is shown in Figure 4.11 for the circles with radius $c = \{15, 16, 17, 18, 19, 21\}$ pixels. When the bone structure is clearly delineated, the best-fitting circle approximates the spinal canal boundary well without ambiguity, and the center of the circle is close to the center of the spinal canal in the corresponding slice, as can be seen in Figure 4.10 (d). The center of the circle with radius 17 pixels is detected as a seed voxel for the spinal canal because the detected center voxel corresponds to the maximum value in the Hough space, as shown in part (c) of Figure 4.11, and satisfies all the imposed constraints.

Figure 4.12 (d) shows the six circles related to the top six values in the Hough space for the CT slice shown in part (a) of the same figure. Figure 4.13 shows slices of the Hough space related to the circles with radius $c = \{13, 14, 20, 21\}$ pixels for the edge map of the vertebral column in Figure 4.12 (c); these slices include all of the top six values in the Hough space. Because of the disjoint structure of the vertebral column, the pixels corresponding to the top three values in the Hough space (marked in green, blue, and cyan, respectively, in Figure 4.13) do not satisfy the selection criteria. As a result, they are rejected and the pixel corresponding to the fourth maximum value (marked in red in Figure 4.13) is selected to be one of the seeds. The fifth (marked in black) and the sixth (marked in yellow) maximum values are shown only for illustrative purposes.

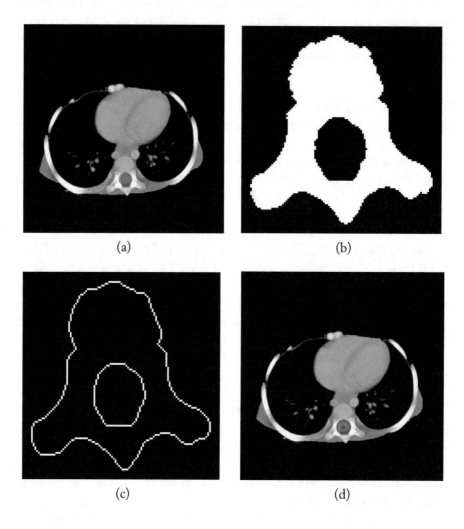

(a)   (b)

(c)   (d)

Figure 4.10: (a) A 512 × 512-pixel CT slice from Exam 1A, with the display range [−300, 400] HU. (b) Detected vertebral column cropped from the slice shown in part (a) and enlarged. (c) Edge map of the detected and cropped vertebral column. (d) The best-fitting circle (shown in red) as determined by the Hough-space analysis. The circle has a radius of 17 pixels or 6.97 mm; the detected center is shown by a red asterisk ("*"). See Figure 4.11 for illustrations of the Hough space for this example. Reproduced with kind permission of Springer Science+Business media and from S Banik, R M Rangayyan, and G S Boag. "Automatic segmentation of the ribs, the vertebral column, and the spinal canal in pediatric computed tomographic images." *Journal of Digital Imaging*, 2009. In press. © SIIM.

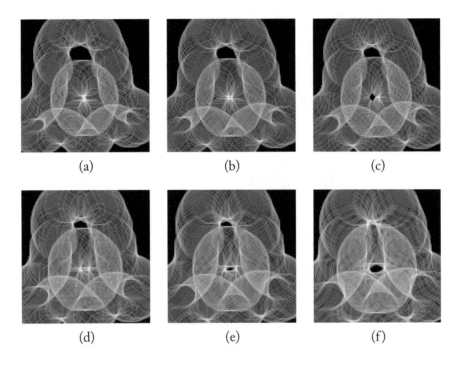

(a)　　　　　　　　(b)　　　　　　　　(c)

(d)　　　　　　　　(e)　　　　　　　　(f)

Figure 4.11:   Illustration of the Hough space for the image in Figure 4.10 (c). (a) for $c = 6.15$ mm (15 pixels), (b) for $c = 6.56$ mm (16 pixels), (c) for $c = 6.97$ mm (17 pixels), (d) for $c = 7.38$ mm (18 pixels), (e) for $c = 7.79$ mm (19 pixels), (f) for $c = 8.61$ mm (21 pixels). The display intensity is $\log_{10}(1 + accumulator\ cell\ value)$ of the Hough space. The detected center is marked by a diamond ('◇') in part (c). The circular paths trace the edges of the vertebral column shown in Figure 4.10 (c). Reproduced with kind permission of Springer Science+Business media and from S Banik, R M Rangayyan, and G S Boag. "Automatic segmentation of the ribs, the vertebral column, and the spinal canal in pediatric computed tomographic images." *Journal of Digital Imaging*, 2009. In press. © SIIM.

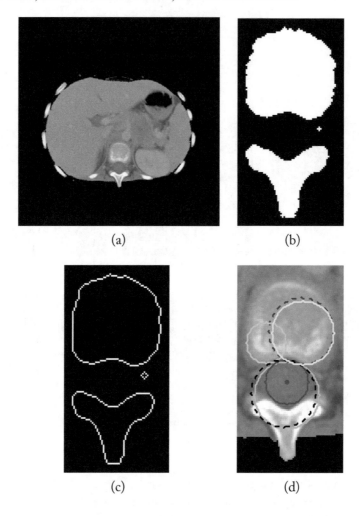

Figure 4.12: (a) A 512 × 512-pixel CT image with the display range [−300, 400] HU. (b) Detected vertebral column cropped from the slice shown in part (a) and enlarged. (c) Edge map of the detected and cropped vertebral column. (d) The green, blue, cyan, red, black, and yellow circles relate to top six values, respectively, in the Hough space. The radius values are: 7.15 mm (13 pixels) for the cyan circle, 7.70 mm (14 pixels) for the red circle, 11.00 mm (20 pixels) for the yellow circle, and 11.55 mm (21 pixels) for the green, blue, and black circles. After satisfying the imposed constraints in the procedure described, the center of the red circle, marked by a dot ('.'), is detected as the seed pixel for the spinal canal. Reproduced with kind permission of Springer Science+Business media and from S Banik, R M Rangayyan, and G S Boag. "Automatic segmentation of the ribs, the vertebral column, and the spinal canal in pediatric computed tomographic images." *Journal of Digital Imaging*, 2009. In press. © SIIM. See Figure 4.13 for illustrations of the Hough space sections for this example.

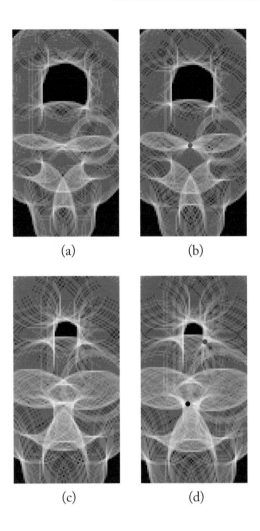

(a)   (b)

(c)   (d)

Figure 4.13: Illustration of the Hough space for the image in Figure 4.12 (c). The top six values in the Hough space are marked with dots ('.') in the same color as the corresponding circles shown in Figure 4.12 (d). (a) Hough parameter space for $c = 7.15$ mm (13 pixels). (b) Hough parameter space for $c = 7.70$ mm (14 pixels). (c) Hough parameter space for $c = 11.00$ mm (20 pixels). (d) Hough parameter space for $c = 11.55$ mm (21 pixels). The display intensity is $\log_{10}(1 + accumulator\ cell\ value)$ of the Hough space. The circular paths trace the edges of the vertebral column and also the artifacts in Figure 4.12 (c). Reproduced with kind permission of Springer Science+Business media and from S Banik, R M Rangayyan, and G S Boag. "Automatic segmentation of the ribs, the vertebral column, and the spinal canal in pediatric computed tomographic images." *Journal of Digital Imaging*, 2009. In press. © SIIM.

4.5.3   EXTRACTION OF THE SPINAL CANAL

The centers of the circles detected by the procedure described in the preceding section are used as the seed voxels for the reconstruction process using the 26-connected neighborhood. The mean and the SD for the reconstruction process are calculated within the neighborhood of 21 × 21 pixels (approximately 1 cm × 1 cm) of each of the seed voxels for the corresponding CT exam. Voxels in the defined neighborhood having HU values not within the range of $23 \pm 3 \times 15$ HU are rejected from the parameter calculation process. Subsequently, the reconstructed fuzzy region is thresholded at $T = 0.80$ and the region is closed in 2D using the convex hull. Then, the result is morphologically closed in 3D using a 3D tubular structuring element of radius 2 mm and height 10 mm. Due to the expected tubular structure of the spinal cord and the spinal column, a tubular structuring element is appropriate [42]. The method described above is summarized in the flowchart in Figure 4.14.

4.5.4   QUALITATIVE EVALUATION OF THE RESULTS

The segmentation methods were applied to 39 CT images of 14 patients (see Section 3.2); Exam 8A was not processed because the exam contains only the pelvic region. The final results of the procedure for three CT exams are shown in 3D in Figure 4.8 as examples, with the corresponding segmented vertebral column shown for reference.

The relatively obvious appearance of the spinal canal within the region enclosed by the vertebral foramen in CT images enables easy visual verification of the results of segmentation. The spinal canal was successfully detected and segmented in all of the CT exams processed. The algorithm fully automates the detection and segmentation of the spinal canal; the method is adaptive as well.

The detected spinal canal in two of the slices from Exam 1C is shown in Figure 4.15. The contours in red correspond to the contours of the spinal canal drawn by the radiologist; the contours shown in dashed green represent the results obtained by the described method. Note that, in Figure 4.15 (a), the contour of the segmented spinal canal matches almost completely with the contour drawn by the radiologist. In Figure 4.15 (b), although the result is in good agreement with the contour drawn by the radiologist, the method could not include regions with low HU values.

4.5.5   QUANTITATIVE EVALUATION OF THE RESULTS

Quantitative comparative analysis of the contours of the spinal canal detected by the segmentation method was performed with reference to the contours drawn independently by a radiologist, for 21 representative CT slices of three patients, by using the measures of Hausdorff distance and the MDCP. The CT slices for the comparative analysis were selected so as to include cervical, thoracic, and lumbar sections with varying shapes of the spinal canal. Several of the selected sections have distinct artifacts due to the partial-volume effect. In most of the slices, the contours obtained by the method described and those drawn by the radiologist overlapped almost completely. The quantitative analysis was performed in the same manner as discussed for the vertebral column in Section 4.4.2. The results of quantitative evaluation are presented in Table 4.2.

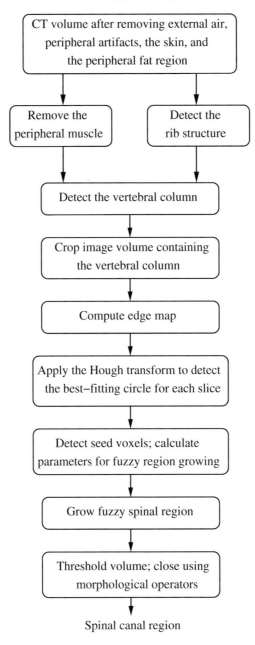

Figure 4.14: Flowchart describing the segmentation of the spinal canal. Reproduced with kind permission of Springer Science+Business media and from S Banik, R M Rangayyan, and G S Boag. "Automatic segmentation of the ribs, the vertebral column, and the spinal canal in pediatric computed tomographic images." *Journal of Digital Imaging*, 2009. In press. © SIIM.

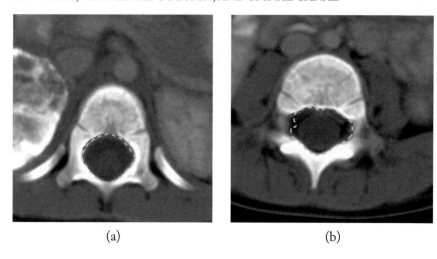

(a)   (b)

Figure 4.15: Two examples of the detected spinal canal from Exam 1C. The contours in red were drawn manually by the radiologist. The contours in green correspond to the boundary of the spinal canal detected by the segmentation procedure. (a) Hausdorff distance = 1.05 mm and MDCP = 0.44 mm. (b) Hausdorff distance = 2.62 mm and MDCP = 0.79 mm. Reproduced with kind permission of Springer Science+Business media and from S Banik, R M Rangayyan, and G S Boag. "Automatic segmentation of the ribs, the vertebral column, and the spinal canal in pediatric computed tomographic images." *Journal of Digital Imaging*, 2009. In press. © SIIM.

The 95% confidence interval, obtained by the one-sample two-tailed t-test, is [1.35, 3.61] for the maximum Hausdorff distances listed in the 4th column of Table 4.2. For the mean MDCP, listed in the 9th column of Table 4.2, the corresponding interval is [0.41, 0.83]. The average MDCP for the three exams processed, listed in Table 4.2, is 0.62 mm, and is approximately equal to the size of one pixel in the CT image. Compared to the representative dimension of the spinal canal in the thoracic region (25 mm) for pediatric patients, the average error (in terms of MDCP) is approximately 2.5%.

The small values of the maximum Hausdorff distance and the maximum MDCP indicate that there is no significant overestimation or underestimation over the 21 slices of three CT exams evaluated. Given that the slices were selected from different sections of the body with varying shapes of the spinal canal, the segmentation method is able to produce good results regardless of the position and shape of the spinal canal; it is also robust, as indicated by the small standard deviation of the results.

Table 4.2: Quantitative evaluation of the results of segmentation of the spinal canal. All distances are in millimeters. Mean = average, SD = standard deviation, min = minimum, and max = maximum. Reproduced with kind permission of Springer Science+Business media and from S Banik, R M Rangayyan, and G S Boag. "Automatic segmentation of the ribs, the vertebral column, and the spinal canal in pediatric computed tomographic images." *Journal of Digital Imaging*, 2009. In press. © SIIM.

Exam	Number of slices	Hausdorff Distance				MDCP			
		min	max	mean	SD	min	max	mean	SD
1C	5	1.05	2.62	1.64	0.63	0.44	0.79	0.62	0.13
2B	10	1.41	2.85	1.92	0.58	0.59	0.86	0.71	0.09
4A	6	0.77	1.97	1.25	0.39	0.35	0.75	0.54	0.14

4.6   APPLICATIONS

The identification and segmentation of the peripheral tissues, the rib structure, the vertebral column, and the spinal canal can facilitate the localization, identification, and segmentation of other organs. The skin can be regarded as the outer-most organ, and can also be used as a landmark. The peripheral fat and the peripheral muscle can interfere with other abdominal tissues in the process of segmentation of several organs as well as tumors; their prior segmentation and removal can lead to better delineation of the regions or organs of interest [11, 12]. In addition, the removal of the skin, the peripheral fat, and the peripheral muscle creates an approach to examine the inner side of the body in a CT image. Furthermore, the information related to the abdominal wall is important in the planning of surgery and recognition of abdominal organs [35]: segmentation of the bone structure and the peripheral muscle structure in CT images can assist in this process.

The rib structure and the vertebral column can be used to introduce relative coordinate systems in the abdomen and the thoracic region to assist in the localization of the corresponding internal organs. The rib structure, the vertebral column, and the spinal canal can be used as important landmarks to perform segmentation, registration, and evaluation of other normal as well as abnormal organs and tissues; they can also be used to assist in the localization of spinal pathology, such as scoliosis and osteoporosis. The methods should find use in procedures for image segmentation, the development of atlases, and in the planning of image-guided surgery and therapy.

In the present work, it has been observed that the abdominal parts of the vertebral column with low HU values interfere with the process of segmentation of the neuroblastic tumor. The spinal canal also creates ambiguity in automatic segmentation of the tumors by increasing the false-positive rates. The 3D rib structure is used in the present work to assist in the procedure for seed selection of the spinal canal (see Section 4.5), and to delimit the lower boundary in the procedure for the segmentation of the lungs (see Section 5.2). The vertebral column and the spinal canal are also used to assist in the process of delineation of the pelvic surface (see Section 6.2). In addition, all of the peripheral artifacts and tissues, the vertebral column, and the spinal canal are removed from consideration in the process of segmentation of the neuroblastic tumor to improve the result of segmentation (see Section 7.3).

4.7 REMARKS

The rib structure, the vertebral column, and the spinal canal can be used as important landmarks for various applications. In this chapter, procedures were described for automatic identification and segmentation of the ribs, the vertebral column, and the spinal canal.

After removing the peripheral artifacts and tissues, the rib structure is identified using morphology-based image processing techniques. Then, the vertebral column is segmented; using the information related to the ribs and the vertebral column, the spinal canal is segmented. The procedure includes several parameters and thresholds that were determined based on experiments with several pediatric CT images.

The results of segmentation were evaluated qualitatively; the results of segmentation of the vertebral column and the spinal canal were also assessed quantitatively by comparing with the results of manual segmentation.

The segmented landmarks have been used to aid the process of delineation of the diaphragm and the pelvic surface, and also in the process of segmentation of the neuroblastic tumor, as described in the subsequent chapters.

CHAPTER 5

Delineation of the Diaphragm

5.1 THE DIAPHRAGM

The diaphragm is a domed-shaped musculofibrous septum that separates the thoracic cavity from the abdominal cavity and performs an important function in respiration [108, 109]. The diaphragm is located inferior to the lungs; the convex upper surface of the diaphragm forms the floor of the thorax on which the lungs and the heart rest [108, 109]. The concave bottom surface of the diaphragm can be considered to be the roof of the abdominal cavity, lying directly over the liver on the right side of the body, and over the spleen and the stomach on the left side [108]. The relative proximity of the diaphragm to several anatomical structures makes it a good reference for landmarking a volumetric medical image.

The diaphragm can be used as an important landmark to aid the delineation of contiguous structures by delimiting the segmentation process of the target organ within the abdomen or the thorax. In the case of segmentation of heterogeneous neuroblastic tumors, leakage of the results has been observed to occur through the heart and into the thoracic cavity [11, 12, 42, 54]. In a related work, Rangayyan et al. [54] proposed a method to identify the diaphragm, aiding in the subsequent removal of the part of the image superior to the diaphragm, that is, the thoracic cavity; the procedure reduced the scope of growth of the tumor segmentation algorithm and led to improved results.

Keatley et al. [110] proposed the use of an augmented active contour model for automated quantification of the motion of the diaphragm in a fluoroscopic movie. Beichel et al. [111] proposed a semi-automatic method for segmentation of the diaphragm; the method involves the use of a 3D active appearance model to segment the surface of the dome of the diaphragm. The approach is limited by the need for extensive user input. Zhou et al. [112] proposed an automatic method for segmentation of the diaphragm using thin-plate splines and obtained good results. However, the exact method of evaluation was not discussed and the procedure is valid only for high-resolution CT data.

The procedures mentioned above may not be suitable for pediatric CT exams because there exist large variations in shape, proportions, and size of the diaphragm in children. Furthermore, the diaphragm may not be readily identifiable in CT images of children due to its small thickness, low spatial resolution of the image data, and the partial-volume effect. The procedure proposed by Rangayyan et al. [44, 54] aims to locate the diaphragm automatically in pediatric CT exams by segmenting the lungs, modeling the base surface as a quadratic surface, and refining it according to a deformable contour model (see Section 2.6). In this work, modifications are made to the methods proposed by Rangayyan et al. [44] in order to improve the definition of the diaphragm. An overview of the modified procedure is given in Figure 5.1, with the details given in the following sections.

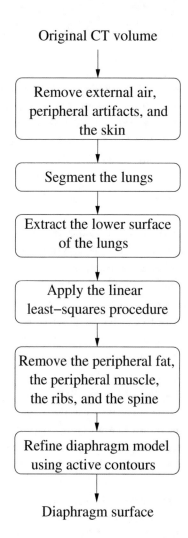

Figure 5.1: Flowchart for the delineation of the diaphragm.

5.2   SEGMENTATION OF THE LUNGS

Several methods have been presented in the literature to perform segmentation of the lungs in CT images. For example, Prasad and Sowmya [113] used a combination of morphological image processing methods; Zrimec and Busayarat [114] used morphological image processing methods in conjunction with active contours to achieve segmentation of the lungs. Brown et al. [115] proposed a method of seeded region growing, edge tracking, and knowledge-based segmentation for the assessment of differential left-lung and right-lung function. Zhou et al. [116] proposed automatic segmentation and recognition of lung structures in high-resolution CT images using the air-filled region in the chest and the airway tree of the bronchi.

Vu [103] and Rangayyan et al. [44] used the procedure proposed by Hu et al. [117] to perform segmentation of the lungs, with a modification to account for the size and proportions of anatomical structures in children. The procedure is based on the fact that the lungs form the single-largest volume of air in the body. In addition, the regions of the lungs comprise mostly low-density tissues (that is, alveoli) and air, making them readily distinguishable from the surrounding structures. It is assumed that there are two types of voxels in the image data: *non-body* voxels comprising the low-density voxels in the lungs and in the air surrounding the body of the patient, and *body* voxels corresponding to voxels within the dense body and chest-wall structures [117]. An optimal thresholding technique is used to separate the body and non-body voxels, and subsequently to identify the lungs. The iterative procedure to determine the optimal threshold can be formulated as follows.

Let T_i be the threshold applied at step i to separate the image data into two regions, and let μ_b and μ_n denote the mean values of the body and non-body regions, respectively, after thresholding with T_i. An updated threshold for step $i + 1$ is computed as

$$T_{i+1} = \frac{\mu_{bi} + \mu_{ni}}{2}, \ i = 1, 2, \ldots . \tag{5.1}$$

The updating procedure is iterated until the optimal threshold value is determined; that is, the algorithm continues until the threshold reaches stability, with $T_{i+1} \approx T_i$. The initial threshold, T_0, is based on the expected CT number for air: $T_0 = -1000$ HU.

After determination of the optimal threshold, the non-body voxels correspond to the air surrounding the body, the lungs, and other low-density regions within the body, such as air in the stomach and bowels. The region surrounding the body is removed using the same procedure as for the removal of external air, the skin, and peripheral artifacts (see Section 4.2). The remaining non-body regions are grouped into disjoint regions, and using binary reconstruction with a 6-connected neighborhood, every voxel is grouped according to its spatially connected neighbors. The volume of each region is determined by counting the connected voxels that comprise the region. Small, disconnected regions are discarded if the corresponding volume is less than 600 voxels each. Regions with volumes greater than 0.5% of the total image voxel count are considered to be the lungs.

In the case of adults, the lungs may be well separated from other low-density structures, such as air in the bowels and stomach; however, due to the small size of organs in children and due to the limited spatial resolution of the CT image data, the air pockets in the bowels and stomach may be erroneously classified as part of the lungs. To avoid this, Vu [103] and Rangayyan et al. [44] proposed a method to examine the cross-sectional (2D) area of the lungs on each slice of the image volume. The lungs are cone-shaped organs with increasing area toward the mid-axial point, where there is a concave cavity to accommodate the heart [108]. From this point downward, the area decreases to a minimum at the lower-most portions of the lungs. Any deviation from this progression, such as an increase in the area near the base of the lungs, could be attributed to regions other than the lungs. Following thresholding, analysis of connected components, and analysis of the cross-sectional area, the segmented lung regions are morphologically closed in 3D, with an ellipsoidal structuring element of size $5 \times 5 \times 3$ voxels, to remove holes present in the result.

Although the methods described above provided good results for the first dataset described in Section 3.2, they failed to produce good results in 13 CT exams in the second dataset by including abdominal air regions (that is, air regions inside the stomach and bowels) in the results, especially with images acquired by the two-breath technique. Because the procedure searches for the largest air region inside the body, and the images acquired in the second breath-hold by the two-breath technique do not contain the entire lungs, the procedure included air regions in the abdomen in 13 CT exams. Considering the fact that the segmented lungs in the present work are used to extract the lower surface of the lungs to perform delineation of the diaphragm, the procedure needs to be modified by concentrating on accurate extraction of the lower surface of the lungs.

To address the problem described above, in the present work, the rib structure is used to restrict the scope of segmentation of the lungs. In the procedure proposed by Vu [103] and Rangayyan et al. [44], the top of the concave cavity between the two lungs is considered to be the slice that is just below the slice containing the largest air region in the body. Here, the ratio between the areas of the air regions in the body and the non-air regions in the body in each slice is taken into account; the slice that corresponds to the maximum of the ratio is considered to be just over the slice of interest.

In order to include the partial lung volume present in the images acquired by the two-breath technique, after following the above-mentioned procedure, the regions with volume greater than 5% of the total image volume within the thoracic region, as determined by the rib structure, are initially considered to be parts of the lungs. Morphological erosion is performed in 2D to split connected regions, and if any connected region in 3D extends below the rib structure, the region is discarded. Then, starting from the bottom of the initially segmented lung structure, the area of each region on each slice is evaluated, moving toward the top of the concave cavity between the two lungs. If any region has a decreasing sectional area toward the top, the corresponding slice is identified. All sections of the region below the identified slice are removed from the result of segmentation.

The results of the modified segmentation procedure are shown in Figures 5.2 and 5.3 for three representative CT exams. The results shown in parts (b) and (d) of Figure 5.2 verify that the procedure is able to produce good results in performing segmentation of the lungs. Figure 5.3 illustrates two examples where the previous methods used by Vu [103] and Rangayyan et al. [44] failed to perform proper segmentation; the modifications described above are able to take care of the deficiencies in the previous methods. Though the modification can provide better results, it fails to remove some of the air regions in the thorax if they are spatially connected to the lungs. The procedure may include the esophagus, trachea, or other small regions with lower HU values than the optimized threshold value; however, such errors do not affect the process of extracting the bottom of the lung surface as well as the delineation of the diaphragm.

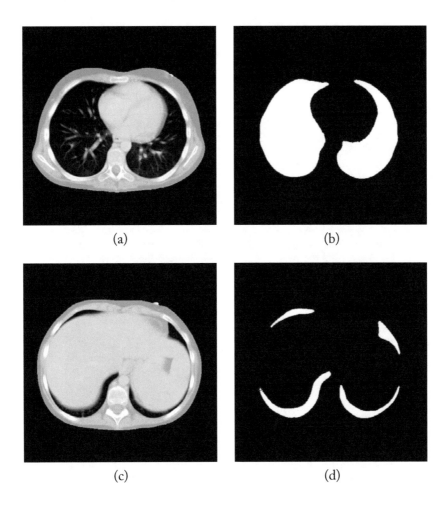

(a)   (b)

(c)   (d)

Figure 5.2: Two 512 × 512-pixel cross-sectional CT slices of Exam 4D: (a) Slice number 24 out of 98. (b) Result of extracting the lung region from (a). (c) Slice number 35. (d) Result of extracting the lung region from (c).

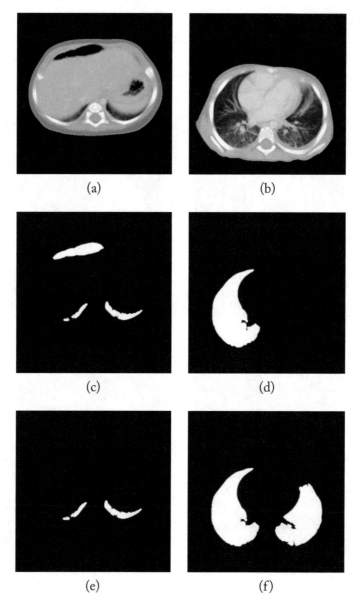

Figure 5.3: Two 512 × 512-pixel cross-sectional CT slices. (a) Exam 3B. Slice 19 out of 48. (b) Exam 5B. Slice 1 out of 57; top of the scan. (c) Result of extracting the lung region from (a) by the previous method; an air pocket in the abdomen has been included in the result. (d) Result of extracting the lung region from (b) by the previous method; the left lung has not been detected. (e) Result of extracting the lung region from (a) by the modified method. (f) Result of extracting the lung region from (b) by the modified method.

5.3   DELINEATION OF THE DIAPHRAGM

The methods proposed by Rangayyan et al. [44, 54] are used to perform the delineation of the diaphragm in the present work with a few modifications. The procedure is described in the following sections.

5.3.1   LINEAR LEAST-SQUARES PROCEDURE TO MODEL THE DI-APHRAGM

After performing the segmentation of the lungs, the initial diaphragm model is taken to be the set of voxels comprising the base of the lungs, where each voxel $\mathbf{v}_i = (x_i, y_i, z_i), i = 1, 2, ..., N$, represents the lowest point (closest to the bottom of the scan), z_i, of the lung surface for a given coordinate pair (x_i, y_i). Here, N is the total number of points in the lower surface of the lungs.

Each dome of the diaphragm is modeled as a quadratic surface, as follows:

$$z_i' = a_0 x_i^2 + a_1 y_i^2 + a_2 x_i y_i + a_3 x_i + a_4 y_i + a_5. \tag{5.2}$$

At each coordinate (x_i, y_i), the error between the estimated diaphragm model and the voxels representing the inferior surface of the lungs is given by $z_i - z_i'$. The error vector is defined as $\mathbf{r} = \mathbf{z} - \mathbf{z}'$, where $\mathbf{z}' = [z_1', z_2', ..., z_N']^T$ and $\mathbf{z} = [z_1, z_2, ..., z_N]^T$, and the squared error is given by $\mathbf{r}^T \mathbf{r}$.

The set of parameters $\hat{\mathbf{a}} = [\hat{a}_0, \hat{a}_1, \hat{a}_2, \hat{a}_3, \hat{a}_4, \hat{a}_5]^T$ that minimizes the squared error is defined as

$$\hat{\mathbf{a}} = \left(\Omega^T \Omega \right)^{-1} \Omega^T \mathbf{z}, \tag{5.3}$$

where

$$\Omega = \begin{bmatrix} x_1^2 & y_1^2 & x_1 y_1 & x_1 & y_1 & 1 \\ x_2^2 & y_2^2 & x_2 y_2 & x_2 & y_2 & 1 \\ x_3^2 & y_3^2 & x_3 y_3 & x_3 & y_3 & 1 \\ \vdots & \vdots & \vdots & \vdots & \vdots & \vdots \\ x_N^2 & y_N^2 & x_N y_N & x_N & y_N & 1 \end{bmatrix}. \tag{5.4}$$

By solving Equation 5.3, the estimated parameters can be used to generate a quadratic surface model for either of the left-dome or the right-dome surface of the diaphragm. The models are combined by calculating the minimum z_i for every coordinate (x_i, y_i) and are restricted to be within the external boundary of the peripheral fat region. The results for the left and right domes are shown separately in Figures 5.4 (a) and (b), for the CT exam shown in Figure 5.2. The combined model, after considering the spatial extent of the diaphragm, is shown in Figure 5.5 (a).

5.3.2   ACTIVE CONTOUR MODELING OF THE DIAPHRAGM

The initial dome structures, obtained by the methods described in the preceding section, can be regarded as a good starting point to delineate the actual diaphragm. To refine the approximation

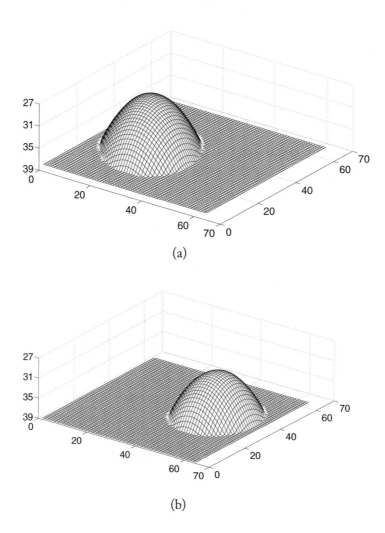

Figure 5.4: Diaphragm model for Exam 4D; see Figure 5.2 for related 2D illustrations. (a) Initial model of the right-dome surface. (b) Initial model of the left-dome surface. See Figure 5.5 for the unified model and the final representation of the diaphragm. The x and y axes are labeled in voxels after down-sampling by a factor of 8, and the z axis represents the slice number.

provided by the least-squares procedure, the active contour model described in Section 2.6 is used to modify the model according to the structural information in the images. On each slice of the image volume, the contour generated by the linear least-squares model is refined to obtain a better segmentation of the diaphragm. For all exams, the following parameters of the deformable contour

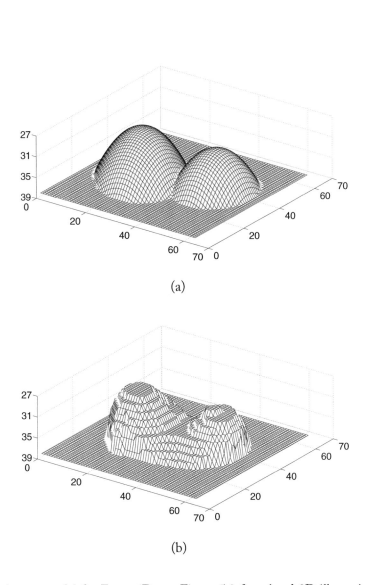

(a)

(b)

Figure 5.5: Diaphragm model for Exam 4D; see Figure 5.2 for related 2D illustrations. (a) Unified model of the diaphragm obtained by combining parts (a) and (b) in Figure 5.4. (b) The result after the application of active contour modeling. The x and y axes are labeled in voxels after down-sampling by a factor of 8, and the z axis represents the slice number.

were used: $\alpha = 30.0$, $\beta = 10.0$, $K_1 = 0.1$, $K_2 = 1.0$, and $\lambda = 0.1$ (see Equations 2.11, 2.15, 2.16, 2.17, and 2.18). The values of the parameters were chosen based on experimentation.

The active contour model converges to align with the nearest sharp edges present in the image; in the present application, the model converged to the edges of the rib structures in several of the cases. In order to prevent such errors and to obtain a better estimation of the diaphragm, in the present work, the peripheral artifacts and tissues, the rib structure, and the vertebral column are removed from the CT images prior to the application of the active contour algorithm.

The results of application of deformable contours to the model obtained by the linear-least squares estimation procedure are shown in part (b) of Figure 5.5, and also in Figures 5.6 and 5.7. Figure 5.6 (a) depicts a CT slice from Exam 4D; the slice is near the top of the diaphragm.

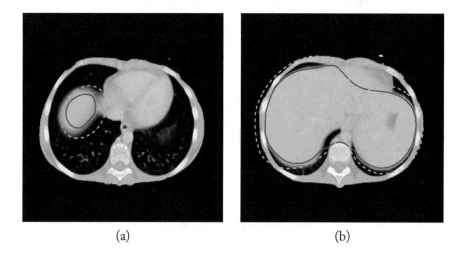

(a)   (b)

Figure 5.6: Two 512×512-pixel CT slices, from Exam 4D. (a) Representation of the diaphragm on CT slice 30 out of 98, near the top of the diaphragm. (b) Representation of the diaphragm on CT slice 35 out of 98, near the bottom of the diaphragm. The initial contour obtained by the linear least-squares procedure is shown in dashed green line and the final representation is shown in red. For consistency, the CT images shown are after the removal of peripheral artifacts, the skin, and the peripheral fat region. See Figure 5.7 for 3D illustrations of the segmented lungs and diaphragm.

The initial approximation of the diaphragm, which is the contour shown by the dashed green line, is an overestimation of the diaphragm. After refinement by the deformable contour model, the final contour obtained is shown in red. Even with a large error in the initial approximation of the diaphragm, as seen in Figure 5.6 (b), applying the deformable contours results in a good definition of the diaphragm, as seen in the same figure. The contour does not fully converge near the vertebral column; however, this results in only a slight overestimation of the diaphragm near the vertebrae.

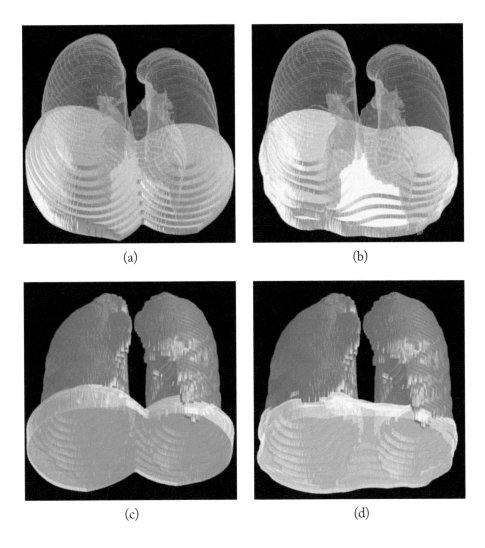

(a)   (b)

(c)   (d)

Figure 5.7: 3D representation of the results of segmentation of the diaphragm of Exam 4D. (a) Initial approximation of the diaphragm (top-down view) (b) Final representation of the diaphragm (top-down view) after applying the active contour model to the approximation in (a). (c) Bottom-up view of the initial approximation of the diaphragm in (a). (d) Bottom-up view of the final, refined representation of the diaphragm in (b). Reproduced with kind permission of Springer Science+Business media and from R M Rangayyan, S Banik, and G S Boag. "Landmarking and segmentation of computed tomographic images of pediatric patients with neuroblastoma." *International Journal of Computer Assisted Radiology and Surgery*, 2009. In press. © Springer. See also Figures 5.4 and 5.5 for related illustrations.

This is a minor problem in the present application because the vertebral column is independently segmented in a preceding step.

The 3D representations of the initial approximation of the diaphragm and the refined representation for Exam 4D are shown in Figure 5.7. The lungs, in blue, are shown for reference. As described in the preceding paragraphs, the initial model of the diaphragm is an overestimation, as seen in Figures 5.7 (a) and (c) as the protrusion of the surface below the lungs. However, following the application of the deformable contours, the final representation of the diaphragm is more accurate and contained within the body, as seen in Figures 5.7 (b) and (d).

5.3.3 QUALITATIVE ASSESSMENT OF THE RESULTS

The methods were applied to 39 CT images of 13 patients (see Section 3.2); Exam 8A was not processed because it contains only the pelvic region. The initial approximation of the diaphragm, as described earlier, is an overestimation of the diaphragm; it encloses the diaphragm, as well as parts of the lungs and the heart. Following the refinements by the deformable contour model, the final contour represents the diaphragm more precisely. Even with increased error in the initial approximation of the diaphragm, application of the deformable contours resulted in a good definition of the diaphragm (see Figure 5.7).

In several CT exams, the contour of the diaphragm did not converge accurately near the vertebral column; however, this results in only a slight overestimation of the diaphragm near the vertebrae, especially close to the bottom of the diaphragm. The results of segmentation of the diaphragm with parts protruding into the bone structures could be an overestimation of the diaphragm, and subsequently, an overestimation of the abdominal cavity in some cases. Without considering the errors related to overestimation as above, the results of segmentation of the diaphragm were observed to be satisfactory for all of the CT exams processed.

After the segmentation of the diaphragm, the results were evaluated on a slice-by-slice basis by comparing the 2D contours with those segmented manually and independently by the radiologist, for 11 CT exams of six patients. Four selected CT slices from Exam 4D are shown in Figure 5.8 as representative cases. In general, the results of the segmentation procedure are in good agreement with the independent segmentation of the diaphragm performed by the radiologist.

In some cases, the procedure for the detection of the diaphragm could produce results which differ from the manual delineation by an experienced observer because of inter-slice partial-volume averaging and poor resolution. In addition, when the initial model of the diaphragm has sections near the vertebral column, the final result could encompass the para-vertebral muscle, or the active contour model may converge to the bone-muscle interface at the vertebrae, leading to an overestimation of the abdominal cavity. All of these phenomena are illustrated in Figure 5.8.

5.3.4 QUANTITATIVE ASSESSMENT OF THE RESULTS

The Hausdorff distance and MDCP measures were used to compare the contour of the diaphragm obtained by the segmentation method for each CT slice, with the corresponding contour drawn

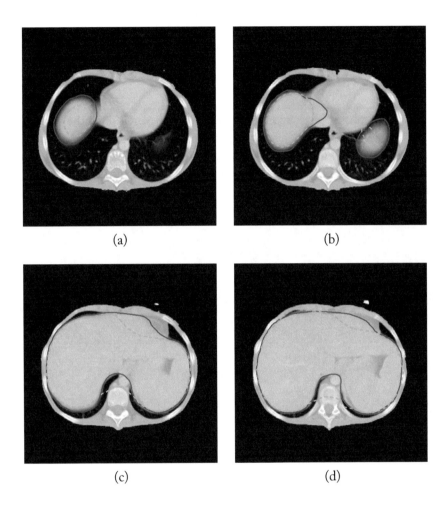

(a)   (b)

(c)   (d)

Figure 5.8: Four 512×512-pixel CT slices from Exam 4D, illustrating qualitative and quantitative analysis of the results of segmentation. The solid red line denotes the contour of the diaphragm drawn independently by the radiologist, and the dashed green line represents the contour of the diaphragm obtained by the method. (a) Slice number 30 out of 98. MDCP = 7.6 mm. Hausdorff distance = 11.7 mm. (b) Slice number 31. Average MDCP = 11.4 mm. Average Hausdorff distance = 17.4 mm. (c) Slice number 35. MDCP = 5.4 mm. Hausdorff distance = 27.7 mm. (d) Slice number 36. MDCP = 5.5 mm. Hausdorff distance = 18.8 mm. For consistency, the CT images are shown after removal of the peripheral artifacts, the skin, and the peripheral fat.

independently by the radiologist, for 11 CT exams of six patients. The results are listed in Table 5.1. The Hausdorff distance was computed for each CT slice where a section of the diaphragm was present, for each CT exam. The minimum, the average, and the SD were computed for each CT exam, across all 2D slices containing a section of the diaphragm. The final Hausdorff distance for the entire diaphragm in 3D was determined by taking the maximum Hausdorff distance over all individual 2D slices for each CT exam.

Table 5.1: Quantitative evaluation of the results of segmentation of the diaphragm. The Hausdorff distance for an entire diaphragm in 3D is the maximum Hausdorff distance over all individual 2D slices containing a section of the diaphragm in the corresponding CT exam. The MDCP for an entire diaphragm in 3D is the average of the MDCP over the related 2D slices in the corresponding CT exam. All distances are in millimeters. Mean = average, SD = standard deviation, min = minimum.

Exam	Number of slices	2D Hausdorff distance section-by-section (mm)			Hausdorff distance over the entire diaphragm in 3D (mm)	MDCP over the entire diaphragm in 3D (mm)
		min	mean	SD		
1A	9	4.54	15.50	5.60	22.31	3.57
1B	9	8.71	13.53	4.36	20.31	3.81
1C	9	9.81	14.30	4.41	21.99	4.53
2B	8	4.68	18.03	8.14	26.25	7.43
3A	7	5.77	8.19	2.33	12.66	3.50
3B	6	6.05	10.08	6.13	22.37	4.67
4A	8	12.58	19.76	4.09	27.94	9.41
4B	9	5.07	20.70	8.05	31.88	7.35
4D	8	5.32	20.62	9.27	34.42	9.41
9B	22	7.02	35.92	15.22	62.19	8.51
14B	14	5.23	17.07	5.93	25.98	4.32
Average					28.03	6.05

The MDCP was calculated on each slice of the dataset where a section of the diaphragm was present, and averaged over each CT exam. If the contours of the right and the left domes of the diaphragm are separated in a particular slice, as can be seen in part (b) of Figure 5.8, the average Hausdorff distance and the average MDCP were calculated for the slice by taking the mean of the corresponding measures for the two contours. Over the 11 CT exams processed of the six patients listed in Table 5.1, a total of 109 CT slices contained sections of the diaphragm. The dataset processed is relatively small because of the difficult and time-consuming task of manual delineation of the contours of the diaphragm on 3D CT exams.

The average error in the results of the procedure described is 28.03 mm, in terms of the average Hausdorff distance in 3D for the CT exams listed in Table 5.1. The overall 3D Hausdorff distance, being the maximum of the Hausdorff distances over all of the related 2D slices, is significantly larger than the corresponding mean Hausdorff distance in all cases because of overestimation or underestimation of the diaphragm in isolated cases of slices. This is indicated by the large deviation in the results for a given exam, as shown in Figure 5.8 and Table 5.1. The method did not perform well for Exam 9B due to poor resolution of the CT image (approximately 0.70 mm per pixel).

On the other hand, because an average of the distances around the contours is used to derive the MDCP, the corresponding result is robust to isolated variations, making MDCP a more reliable indicator of the accuracy and success of the method.

The 95% confidence interval, obtained by the one-sample two-tailed t-test, is [19.46, 36.60] for the maximum Hausdorff distances listed in the 6th column of Table 5.1. For the mean MDCP, listed in the 7th column of Table 5.1, the corresponding interval is [4.44, 7.65]. The average MDCP for the 11 CT exams processed, listed in Table 5.1, is 6.05 mm; compared to the representative dimension of the diaphragm (170 mm) for pediatric patients, the error is approximately 3.6%.

The effects of removal of the rib structure and the vertebral column prior to the application of the active contour model are illustrated in Figure 5.9. The contour obtained by the modified method is shown in yellow. The results of the modification have not included the rib structure near the edges. In comparison with the results reported by Rangayyan et al. [44] using the first nine CT exams listed in Table 5.1, the average MDCP was reduced from 5.85 mm to 5.71 mm and the average Hausdorff distance was reduced from 24.6 mm to 24.4 mm as a result of the modification.

5.4   APPLICATIONS

Segmentation of the diaphragm provides an approach to restrict and constrain the scope of algorithms for segmentation of thoracic or abdominal organs and tumors, and serves as a potential landmark in the delineation of contiguous structures. For the purpose of reducing the false-positive error rate in the segmentation of abdominal neuroblastic tumors, the procedure for automatic detection of the diaphragm could be effective in limiting the tissue volume to be processed, with low possibilities of excluding portions of the primary tumor (see Chapter 7).

The effect of removing the thoracic cavity from the CT exam shown in Figure 5.6, 5.7, and 5.8 is illustrated in Figure 5.10; the lungs are shown for reference. By generating and refining the diaphragm model, and subsequently removing the thoracic cavity, there no longer exists the potential for the results of tumor segmentation to leak into the thorax; this should improve the results of tumor segmentation.

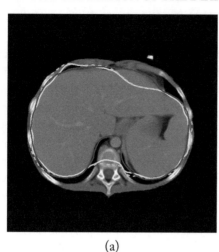 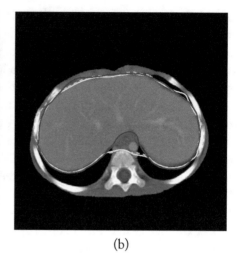

(a)   (b)

Figure 5.9: Two 512×512-pixel CT slices to demonstrate the effect of the modification in the process of delineation of the diaphragm. The contours obtained by the previous procedure are shown by dashed cyan lines; the results after the modification are shown by solid yellow lines. The solid red lines represent the corresponding contours drawn by the radiologist. (a) Representation of the diaphragm on CT slice 36 out of 98, near the bottom of the diaphragm in Exam 4D. Corresponding MDCP reduced from 4.0 mm to 3.4 mm and the Hausdorff distance reduced from 20.3 mm to 18.7 mm. (b) Representation of the diaphragm on CT slice 24 out of 75, near the bottom of the diaphragm in Exam 1A. Corresponding MDCP reduced from 2.3 mm to 1.7 mm and the Hausdorff distance reduced from 19.7 mm to 18.7 mm. Note that, due to the removal of the rib structure from the CT exams, the contours obtained by the method have not converged to the outer boundary of the rib structure. For consistency, the CT images are shown after the removal of peripheral artifacts, the skin, and the peripheral fat region, and without removal of the vertebral column and the rib structure.

5.5   REMARKS

A method to segment the diaphragm automatically based on the spatial proximity of the muscle to the lungs was described in this chapter. The method developed in the present work includes modifications and improvements to a previously developed method. The procedure includes several parameters and thresholds that were determined based on experiments with several pediatric CT images.

The method was applied to 39 CT images with poor interslice resolution; in general, the results were observed to be good. It could be expected that the method will provide higher accuracy with CT images possessing higher spatial resolution, lower interslice distance, and lower error because of the partial-volume effect.

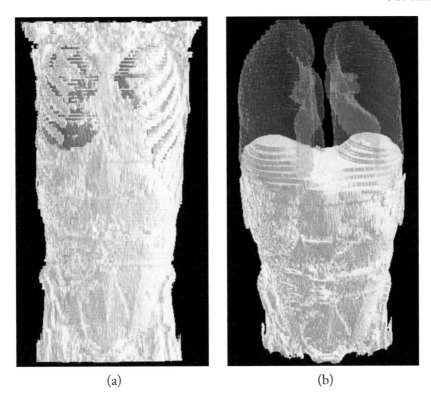

(a)   (b)

Figure 5.10: Illustrations to demonstrate the removal of the thoracic cavity using the diaphragm surface obtained. The 3D representations of the body shown (Exam 4D) are after removal of the external air, peripheral artifacts, the skin, the peripheral fat, and the peripheral muscle. (a) Before removal of the thoracic cavity. (b) After removal of the thoracic cavity. The lungs are shown, in transparent blue, for reference. See Figure 4.3 for illustrations of other surfaces of the same CT exam.

Although the overestimation of the abdominal cavity encountered in some CT slices is favorable in the application of segmenting abdominal tumor masses, it is necessary to refine the algorithm further to improve the accuracy of identification of the diaphragm near the vertebrae. Further work is also required to determine the optimal parameters of the deformable contour model used.

The segmented diaphragm can be used as a landmark to localize and segment abdominal or thoracic organs. In addition, it can be used as a separating boundary between the thorax and the abdomen, and can aid in restricting region growing or other image processing algorithms within the corresponding expected portion of the body. Use of the diaphragm in the problem of segmentation of neuroblastic tumors is discussed in Chapter 7.

CHAPTER 6

Delineation of the Pelvic Girdle

6.1 THE PELVIC GIRDLE

The *bony pelvis* or the *pelvic girdle*, also known as the *hip girdle*, is composed of two hip bones [118]. During childhood, each hip bone consists of three separate parts: the ilium, the ischium, and the pubis. In an adult, these three bones are firmly fused into a single bone. At the back, the two hip bones meet on either side of the sacrum, and at the front, they are connected at the pubic symphysis [118, 119].

The pelvic girdle serves several important functions in the body. It supports the weight of the body from the vertebral column. It also protects and supports the lower organs, including the urinary bladder and the reproductive organs, as well as the developing fetus in a pregnant woman. The pelvic girdle differs between men and women. In a man, the pelvis is more massive and the iliac crests are closer together, whereas in a woman, the pelvis is more delicate and the iliac crests are farther apart. These differences relate to the woman's role in pregnancy and delivery of children [118].

Digital atlases are widely used for registration, localization, and segmentation of organs and structures in the body as well as for assistance in the planning of surgery [30, 31, 33, 34]. However, clinical cases are often complicated and vary greatly. As a result, few reliable landmarks may exist in pediatric CT images to aid registration or segmentation of organs or tumors, or to facilitate the use of standard atlases. Identification and segmentation of the pelvic girdle can be helpful in this regard.

The identification and removal of the lower pelvic region could also prevent some of the leakage in the result of segmentation of the neuroblastic tumor and reduce the false-positive error rate [54, 55]. However, in pediatric subjects, the pelvic girdle may not be fully developed, both in structure and tissue characteristics. There could also be large variations in the shape, proportions, and size of the pelvis in children. These factors create difficulties in segmentation of the pelvic girdle in CT exams of children.

Automated segmentation of the pelvis in CT images is a challenging task, due to several distinct difficulties, such as nonuniformity of bone tissue, presence of blood vessels, diffused and weak edges, narrow inter-bone regions, poor resolution of CT images, and the partial-volume effect [120]. A few works have been reported on the segmentation and registration of the pelvis. Zoroofi et al. [120] proposed a method for segmentation of the pelvis and femur by an adaptive thresholding technique based on neural networks. Ellingsen and Prince [121] used deformable registration methods to perform segmentation of the pelvis. Automatic segmentation of the hip bone using nonrigid registration was presented by Pettersson et al. [122]. Westin et al. [123] developed a tensor-based filter for enhancement of CT images and showed that use of the filter in the preprocessing stage is effective for improving the result of a conventional thresholding method for segmentation of the pelvis and the femur.

All of the works cited above are applicable to CT exams of adults, and are not appropriate for application to pediatric CT images. In the present chapter, methods are described to detect, segment, and use the pelvic girdle and the upper pelvic surface as landmarks to aid the identification of lower abdominal organs, and to separate the abdominal cavity from the lower portion of the body.

6.2   DELINEATION OF THE PELVIC GIRDLE

Each CT volume is processed to identify and remove peripheral artifacts, the skin, the peripheral fat, the peripheral muscle, the vertebral column, and the spinal canal using the procedures described in Chapter 4. By applying reconstruction with fuzzy mapping, the pelvic girdle is identified and segmented using the following procedure.

6.2.1   DETECTION OF SEED VOXELS IN THE PELVIC GIRDLE

The vertebral column and the spinal canal segmented by the procedures described in Chapter 4 are used as landmarks for automatic selection of seed voxels in the pelvic girdle. The lowest point of the spinal canal region is taken as the reference, and the CT slice one section (2.5 or 5 mm) away from the reference slice toward the thorax is considered for selection of the seeds. The procedure for automatic seed detection is described as follows.

1. To prevent the seed voxels from being located in a region other than the ilia, such as in the sacrum or the fifth lumbar vertebra, the vertebral column is removed prior to seed detection.

2. To obtain appropriate seed voxels in the left and the right hip bones, a search is performed for high-density regions in the inward direction from the inner boundary of the peripheral fat region, along the line through the center of the spinal canal and parallel to the x-axis (see Figure 6.1).

3. The high-density voxels on the line specified as above, having values higher than 400 HU, and located within 3 cm of the inner boundary of the peripheral fat region, are considered for seed selection.

4. If no voxel on the specified line satisfies the constraints as above, the search is performed in the same manner by moving from the specified line, parallel to the x-axis, toward the front (that is, anterior) of the body, in the same slice (see Figure 6.1).

5. The voxels located in the outer portion of the pelvic girdle (that is, close to the inner boundary of the peripheral fat region) are considered as the detected seed region.

6. The seed selection procedure as above is performed separately for the left and the right hip bones.

The procedure is illustrated in Figure 6.1. If regions 'A' and 'B', as shown in the figure, are not connected, then the separation between the two regions is examined; if the separation is larger than 1.5 mm, region 'B' is rejected. This procedure is required to obtain the seed voxels in the outer portion of the ilium and to prevent the detected seed voxels from being located on the spine or any other unwanted region. The seed pixels obtained by the procedure as above, for both the right and the left hip bones, are used as region markers in the reconstruction process, as described in the following sections.

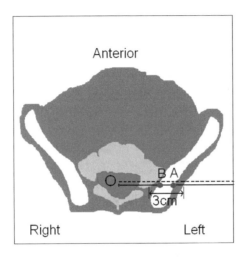

Figure 6.1:   Schematic representation of the seed search procedure for the left side of the pelvic girdle. The search is performed up to 3 cm inward from the inner boundary of the peripheral fat region. The high-density regions on the specified line are marked as 'A' and 'B', and the center of the spinal canal region is denoted by 'O'. If the separation between 'A' and 'B', is larger than 1.5 mm, 'B' is rejected and the voxels in the region 'A' are selected as the seeds. If no voxel on the specified line satisfies the given criteria, the search is performed by moving toward the anterior of the body on the same slice (dotted line). The vertebral column is removed prior to seed detection, and is shown in light gray for reference. The region shown in blue represents the segmented spinal canal. Reproduced with permission from S Banik, R M Rangayyan, and G S Boag. "Delineation of the pelvic girdle in computed tomographic images". In *Proceedings of the 21st IEEE Canadian Conference on Electrical and Computer Engineering*, pages 179–182, Niagara Falls, Ontario, Canada, May 2008. © [2008]IEEE.

6.2.2   SEGMENTATION OF THE PELVIC GIRDLE

To perform the segmentation of the pelvic girdle, the preprocessed CT image is mapped using a fuzzy membership function with the mean $\mu = 400$ HU and standard deviation $\sigma = 120$ HU. These values were determined by manual estimation of the pelvic region, including bone and bone marrow, using 10 pediatric CT exams. The selected seeds, determined by the procedure described above, are used independently as region markers to perform reconstruction with a 26-connected neighborhood in 3D to obtain the left and the right hip bones.

The region above the pelvic girdle is removed before performing reconstruction to prevent leakage into the upper abdominal organs and the vertebral column. For this purpose, the upper limit of the pelvic girdle (that is, the iliac crest) is found by searching for high-density regions within 3.5 cm inward from the inner boundary of the peripheral fat region; the search is performed in a manner similar to that for seed selection. Starting from the slice containing the lower end of the spinal canal, the process is repeated for each slice toward the thorax, until the slice in which no pixel can be found to satisfy the inclusion criteria is reached.

The image resulting after reconstruction is morphologically closed using a disk-type structuring element of radius 5 pixels (approximately 2.5 mm), and thresholded to the range of 200 HU to 1200 HU. After filling the holes in the image, the identified pelvic region is examined, and the connected objects (in 3D) other than the largest two are rejected to remove small isolated regions.

6.2.3   LINEAR LEAST-SQUARES PROCEDURE TO MODEL THE UPPER PELVIC SURFACE

To obtain the upper surface of the pelvic region, the linear least-squares procedure described in Section 5.3.1 to model the diaphragm is applied, with a modification to extract the upper surface of the pelvic girdle. In the case of diaphragm, the bottom of the lung surface is used as a reference; in the case of the pelvic girdle, an initial model obtained by the linear least-squares estimation procedure is defined using the set of voxels comprising the upper surface of the segmented pelvic girdle, where each voxel $\mathbf{v}_i = (x_i, y_i, z_i), i = 1, 2, ..., N$, represents the topmost point (closest to the head), z_i, of the pelvic surface for a given coordinate pair (x_i, y_i). Here, N is the number of points in the upper surface of the pelvic girdle.

By solving Equation 5.3, the estimated parameters can be used to generate a quadratic model for the upper pelvic surface. The inner boundary of the peripheral fat region is used as the limit of the outer boundary of the modeled surface. The modeled pelvic surface is shifted by 1 cm toward the abdomen to facilitate better separation of the abdominal cavity.

6.2.4   ACTIVE CONTOUR MODELING OF THE UPPER PELVIC SURFACE

The pelvic surface, obtained as above, is refined by using the deformable or active contour model in conjunction with the GVF method as described in Section 2.6. The active contour with GVF is initialized with the result of the linear least-squares estimation procedure, and is applied on a slice-by-slice basis to the CT volume to obtain the final representation of the upper pelvic surface. The same parameters as mentioned in Section 5.3.2 are used for the process.

6.2.5   QUALITATIVE ASSESSMENT OF THE RESULTS

The procedure for delineation of the pelvic girdle was tested on 39 CT exams of 14 patients (see Section 3.2). Exam 11A was not processed because it contains only the thoracic region.

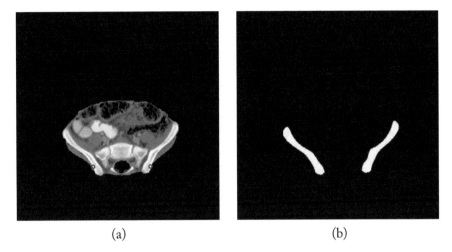

        (a)   (b)

Figure 6.2: (a) A 512 × 512-pixel CT slice from a Exam 6C to illustrate the results of segmentation of the pelvic girdle. The seed voxels selected on the left and the right hip bones are marked in red. (b) Detected parts of the right and left hip bones. The image in part (a) is shown after removal of peripheral artifacts, the skin, the peripheral fat, and the peripheral muscle. The vertebra is shown for reference in part (a).

Representative results of segmentation of the pelvic girdle are shown in Figures 6.2 and 6.3 for three CT slices taken from three different CT exams. In part (a) of Figure 6.2, the seed voxels detected by the procedure are marked in red. Note that the areas of bone marrow having lower HU values are not included in the result. The method performed well even in the presence of neighboring contiguous structures with similar HU values as shown in Figure 6.2 (b) and Figure 6.3 (b). The method failed to include precisely all the parts of the lower pelvic region for the image displayed in part (c), as shown in part (d) of Figure 6.3. Due to the partial-volume effect, some parts in the pubic bones have lower HU values, and are not included in the result; the femoral heads are included in the result, as can be seen in Figure 6.3 (d).

The refined surface obtained by the described procedure was observed to be a good representation of the upper surface of the pelvic girdle. The convergence of the active contours depends largely on the initialization provided by the linear least-squares model and also on the image itself. In three CT exams, the linear least-squares model failed to provide proper initialization because of the presence of other lower abdominal structures and improper positioning of the patient. In Exam 8A, the scan contains a pelvic tumor, and the method could not function appropriately without including the tumor.

However, the segmentation method successfully produced good representation of the pelvic surface in all the other exams, and can be used to separate the lower pelvic region from the abdominal

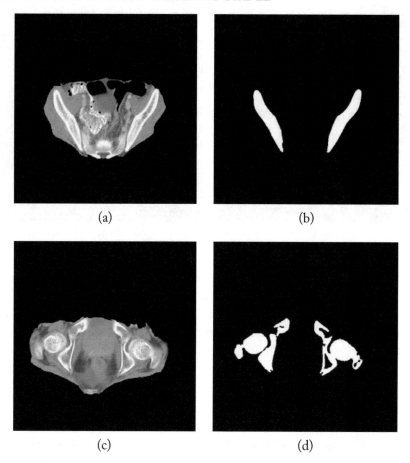

(a)   (b)

(c)   (d)

Figure 6.3: Two 512×512-pixel CT slices from two different CT exams to illustrate the results of segmentation of the pelvic girdle. (a) A CT slice from Exam 4A. (b) Detected parts of the right and the left hip bones for the image shown in part (a). (c) A CT slice from Exam 4D. (d) Detected parts of the pelvic girdle for the image shown in part (c). The procedure failed to extract precisely all the parts in the lower pelvic region. The selected seeds are not located in the slices shown for the two exams in parts (a) and (c). The images in parts (a) and (c) are shown after removal of peripheral artifacts, the skin, the peripheral fat, and the peripheral muscle. The vertebrae is shown for reference in part (a).

cavity. 3D representations of the linear least-squares model and the refined pelvic surface for a CT exam are illustrated in Figure 6.4. The segmented pelvic girdle is shown for reference for the corresponding CT exams. Note that the obtained surface follows the actual structure of the pelvic girdle.

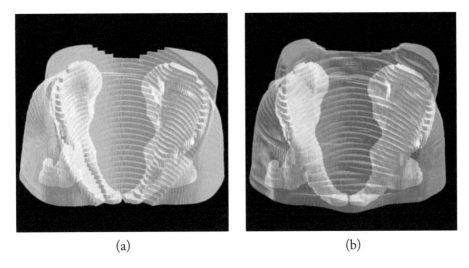

(a)   (b)

Figure 6.4:   Examples to demonstrate the results of the procedure to model the pelvic surface. The pelvic girdle segmented by the segmentation method is shown for reference. (a) Model of the upper pelvic surface (in Exam 4B) obtained by the linear least-squares estimation. (b) The upper pelvic surface after application of the active contour model for the same CT exam. Reproduced with permission from S Banik, R M Rangayyan, and G S Boag. "Delineation of the pelvic girdle in computed tomographic images". In *Proceedings of the 21st IEEE Canadian Conference on Electrical and Computer Engineering*, pages 179–182, Niagara Falls, Ontario, Canada, May 2008. © [2008] IEEE.

After the segmentation of the pelvic girdle, the results were evaluated on a slice-by-slice basis by comparing the 2D contours with those segmented manually and independently by the radiologist, for 13 CT exams of six patients. Four selected CT slices from four different patients are shown in Figure 6.5 as representative cases. In general, the results of the procedure were observed to be in good agreement with the independent segmentation of the pelvic girdle by the radiologist. In all the cases evaluated, both the contours overlapped completely, except in the lower pelvic region. Due to the disjoint bone structures in the lower pelvic region, the error was observed to be high for the corresponding contours.

Given that the aim in the present work is to obtain the upper surface of the pelvic girdle, the results of segmentation were observed to be good representations of the pelvic girdle. Because of the removal of the vertebral column prior to the implementation of the procedure, parts of the lower spinal region (that is, the lumbar vertebra, the sacrum, and the coccyx) are not included in the result, except in two CT exams, where the parts of the corresponding vertebral column were not present in the result of segmentation of the vertebral column because of fusion with the hip bones.

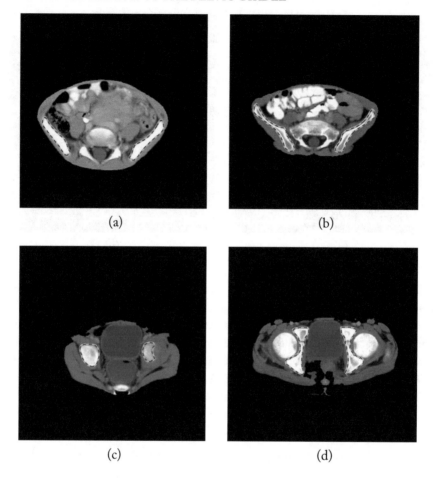

(a)   (b)

(c)   (d)

Figure 6.5:   Four 512×512-pixel CT slices, from four different patients, illustrating qualitative and quantitative analysis of the results of segmentation. The solid red line denotes the contour of the pelvic girdle section drawn independently by the radiologist, and the dashed green line represents the contour of the pelvic girdle produced by the segmentation method. (a) A slice from Exam 7B. Average MDCP = 0.4 mm. Average Hausdorff distance = 1.6 mm. (b) A slice from Exam 5E. Average MDCP = 0.4 mm. Average Hausdorff distance = 2.7 mm. (c) A slice from Exam 6B. Average MDCP = 0.5 mm. Average Hausdorff distance = 1.3 mm. (d) A slice from Exam 13A. Average MDCP = 0.5 mm. Average Hausdorff distance = 1.7 mm.

In five CT exams, the procedure failed to include the lower pelvic girdle region (that is, the ischium and/or the pubis) precisely, due to the disjoint appearance of the bony pelvis in pediatric CT exams and low spatial resolution. Due to the fact that the bones in the ischium and the pubis are not fused together in children, and may exhibit low HU values in pediatric CT exams, it is reasonable that the method cannot include them in the result. In seven CT exams, in some slices, the bone marrow was not included in the result of segmentation. However, the outer portions of the bones were correctly extracted, and as a result, the boundaries of the pelvic girdle in the corresponding slices were properly obtained. In three CT exams, the upper limit of the pelvis was not detected exactly due to the presence of tumors and structures with high HU values in close proximity to the iliac crest; as a result, the structures mentioned were spatially connected to the actual pelvic girdle.

In addition to poor spatial resolution and similar CT characteristics of neighboring regions, due to the inclusion of the contrast material in the abdominal organs, some of the structures show higher HU values than normal, and were included in the results. In five CT exams, some structures in the lower abdomen with similar CT characteristics and with spatial connection to the pelvic girdle were included in the results of segmentation. In addition, the acetabulum and the femoral heads were included in the result in several cases.

6.2.6   QUANTITATIVE ASSESSMENT OF THE RESULTS

The Hausdorff distance and MDCP measures were used to compare the contours of the pelvic girdle obtained by the method for each CT slice, with the corresponding contours drawn independently by the radiologist, for 13 CT exams of six patients. The results are listed in Table 6.1.

The Hausdorff distance was computed for each CT slice where a section of the pelvic girdle was present for each CT exam. The minimum, average, and SD were computed for each CT exam, across all 2D slices containing a section of the pelvic girdle. The final Hausdorff distance for the entire pelvic girdle in 3D was determined by taking the maximum Hausdorff distance over all individual 2D slices containing the pelvic girdle for each CT exam. The average Hausdorff distance for each slice was obtained by taking the mean of the individual Hausdorff distances for the left and the right sides of the pelvic girdle.

The MDCP was calculated on each slice of the dataset, where a section of the pelvic girdle was present, and averaged over each CT exam. The average MDCP for each slice was obtained by taking the mean of the individual MDCP for the left and the right sides of the pelvic girdle. Over the 13 CT exams processed of six patients, as shown in Table 6.1, a total of 277 CT slices contained sections of the pelvic girdle.

The average error in the results of the procedure is 5.95 mm, in terms of the average Hausdorff distance in 3D for the 13 CT exams listed in Table 6.1. However, the overall 3D Hausdorff distance, being the maximum of the Hausdorff distances over all of the related 2D slices, is significantly larger than the corresponding mean Hausdorff distance in all cases because of overestimation or underestimation of the pelvic girdle in isolated cases of slices, especially in the lower pelvic region. Otherwise, the contours of the detected pelvic girdle and the corresponding contours drawn manually overlapped completely and introduced negligible error.

On the other hand, because an average of the distances around the contours is used to derive the MDCP, the corresponding result is robust to isolated variations, making MDCP a more reliable indicator of the accuracy and success of the segmentation method than the Hausdorff distance.

Table 6.1: Quantitative evaluation of the results of segmentation of the pelvic girdle. The Hausdorff distance for an entire pelvic girdle in 3D is the maximum Hausdorff distance over all individual 2D slices containing a section of the pelvic girdle in the corresponding CT exam. The MDCP for an entire pelvic girdle in 3D is the average of the MDCP over the related 2D slices in the corresponding CT exam. All distances are in millimeters. Mean = average, SD = standard deviation, and min = minimum. Reproduced with permission from S Banik, R M Rangayyan, G S Boag, and R H Vu. "Segmentation of the Pelvic Girdle in Pediatric Computed Tomographic Images." Submitted to the *Journal of Electronic Imaging*, June 2008. © SPIE.

Exam	Number of slices	2D Hausdorff distance section-by-section (mm)			Hausdorff distance over the entire pelvic girdle in 3D (mm)	MDCP over the entire pelvic girdle in 3D (mm)
		min	mean	SD		
5B	20	0.86	1.96	0.93	4.75	0.59
5C	21	0.99	2.27	1.27	7.11	0.58
5D	19	1.09	2.45	1.21	5.57	0.55
5E	20	0.97	2.17	1.04	4.87	0.52
6A	16	0.98	1.76	0.87	4.40	0.51
6B	17	0.71	1.99	0.92	4.99	0.50
6C	17	0.98	2.18	1.13	6.11	0.50
7B	18	0.96	1.98	0.97	4.91	0.57
7C	17	0.96	2.09	1.00	4.31	0.57
7D	17	0.86	2.12	1.12	4.30	0.57
12B	28	0.68	1.69	0.13	6.36	0.48
13A	29	0.82	2.53	1.28	5.85	0.59
14B	38	0.66	2.08	1.85	13.85	0.34
Average					5.95	0.53

The 95% confidence interval, obtained by the one-sample two-tailed t-test, is [4.43, 7.48] for the maximum Hausdorff distances listed in the 6th column of Table 6.1. For the mean MDCP, listed in the 7th column of Table 6.1, the corresponding interval is [0.49, 0.57]. The average MDCP for the 13 CT exams processed, listed in Table 6.1, is 0.53 mm, and is comparable to the size of one pixel in a CT image. Compared to the representative dimension of the ilia (85 mm) for pediatric patients, the error is approximately 0.6%.

The quantitative evaluation (as above) was performed only for the segmented pelvic girdle; the pelvic surface obtained by the modeling procedure was not evaluated objectively. Although it is possible to perform quantitative comparison of the pelvic surface obtained by modeling with the corresponding surface derived from the manually drawn contours, such analysis is not required for the present application, which is to facilitate the separation of the pelvic cavity from the abdominal cavity.

6.3   APPLICATIONS

The modeling of the pelvic surface helps to separate the abdominal cavity for further consideration; in addition, the segmented pelvic girdle may be used as a landmark for identification of abdominal organs. Although methods have been described to represent only the upper surface of the pelvic girdle, it should be possible to extend the methods to delineate the lower pelvic region to aid the identification of lower pelvic organs, and also to separate the pelvic cavity for other purposes.

In the present work, the method of delineation of the pelvic girdle is used to reduce the false-positive error rate in segmentation of abdominal tumors due to neuroblastoma, as presented in Chapter 7. The method prevents leakage of the result of segmentation of the tumor into the lower region of the pelvis, and restricts the tumor segmentation process to the abdominal cavity. Figure 6.6 illustrates a CT exam in 3D, including the separation of the abdominal and pelvic regions by the detected upper surface of the pelvic girdle. Figure 6.7 illustrates several important landmarks segmented by the segmentation and modeling methods presented in the preceding chapters. In part (a) of Figure 6.7, the lungs, the upper surface of the abdomen, and the upper surface of the pelvic girdle are shown. Some of the segmented landmarks, such as the rib structure, the vertebral column, the spinal canal, the diaphragm, and the pelvic girdle are shown in parts (b) and (c) of the same figure.

6.4   REMARKS

A method to perform segmentation of the pelvic girdle automatically and to use the segmented pelvic girdle to obtain the upper pelvic surface is described in this chapter. The procedure includes parameters and thresholds that were determined based on experiments with several pediatric CT images.

The methods were applied to 39 CT images; in general, the results were observed to be good. The segmented pelvic girdle can be used as a landmark to localize and segment abdominal or pelvic organs. In addition, the pelvic surface obtained by the method can be used as a separating boundary between the lower pelvic region and the abdomen, and could aid in restricting region growing or other image processing algorithms within the expected portion of the body. The upper pelvic surface can also be useful to localize the lower pelvic organs.

Although the overestimation of the pelvic surface encountered in some CT slices is favorable in the application of segmenting abdominal tumor masses, it is desirable to refine the algorithm

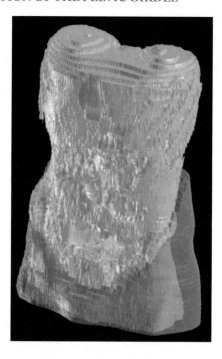

Figure 6.6: An example (Exam 4A) to demonstrate the application of the upper surface of the pelvic girdle obtained by the modeling procedure to separate the abdomen from the lower pelvic regions. A 3D representation of the body is shown after removing the external air, peripheral artifacts, the skin, the peripheral fat, the peripheral muscle, and the thoracic cavity. The pelvic surface obtained is shown in magenta.

further to improve the accuracy of the upper pelvic surface. Quantitative assessment of the pelvic surface obtained by modeling is desirable. Further work is also required to determine the optimal parameters of the deformable contour model used.

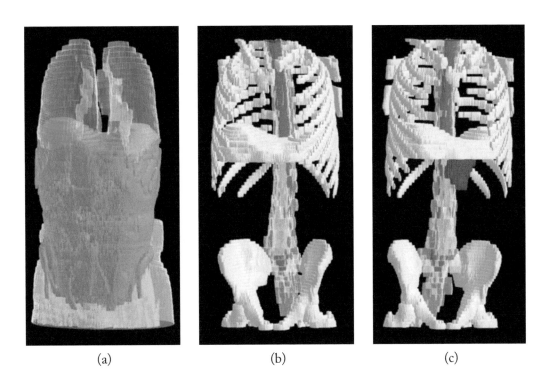

(a)   (b)   (c)

Figure 6.7:   Examples to illustrate some of the segmented landmarks in a CT exam (Exam 4D). (a) A 3D representation of the body after removal of the external air, peripheral artifacts, the skin, the peripheral fat, the peripheral muscle, and the thoracic cavity. The upper surface of the pelvic girdle is shown in transparent yellow and the lungs are shown in transparent blue. (b) 3D representations of the rib structure (cyan), the vertebral column (gray), the spinal canal (light red), the diaphragm (yellow), and the pelvic girdle (cyan). (c) The tumor segmented by the radiologist is shown in red along with several other landmarks. Reproduced with kind permission of Springer Science+Business media and from R M Rangayyan, S Banik, and G S Boag. "Landmarking and segmentation of computed tomographic images of pediatric patients with neuroblastoma." *International Journal of Computer Assisted Radiology and Surgery*, 2009. In press. © Springer.

CHAPTER 7

Application of Landmarking: Segmentation of Neuroblastoma

7.1 NEUROBLASTOMA

Neuroblastoma is the most common extra-cranial solid malignant tumor in children [13, 51, 52, 53]; it is the third most common malignancy of childhood, surpassed in incidence only by acute leukemia and primary brain tumors [124, 125]. Neuroblastoma accounts for 8 − 10% of all childhood cancers, but is responsible for 15% of all cancer-related deaths in the pediatric age group [10, 50, 126]. About 65% of neuroblastic tumors are located in the abdomen; approximately two-thirds of these arise in the adrenal gland. 15% of the tumors are thoracic, usually located in the sympathetic ganglia of the posterior mediastinum. About 10 − 12% of neuroblastic tumors are disseminated without a known site of origin [124].

7.2 COMPUTER-AIDED ANALYSIS OF NEUROBLASTOMA

In the treatment of patients with neuroblastoma, the ultimate goal or the treatment of choice is the complete surgical resection of the tumor mass [53]. However, due to the size or extension of the mass, radiation therapy or chemotherapy may first be required to shrink the tumor before resection can be performed. As such, radiological evaluation of the tumor could provide important indications of the response of the disease to therapy; advances in radiology have made possible the detection and staging of the disease [10, 127]. In this context, computer-aided analysis in the form of tumor segmentation and analysis [11, 12, 17] can be beneficial to radiologists, providing for quantitative and reproducible evaluation of the tumor. Segmentation [11, 12, 128] and analysis of the tissue composition of the tumor [17] could assist in quantitative assessment of the response to therapy and in the planning of delayed surgery for resection of the tumor.

The difficulty in segmentation is caused by the heterogeneity of the tumor mass: neuroblastic tumors generally comprise several different tissues with varying density, including necrosis (low density), viable tumor (medium density), and calcification (high density); furthermore, several contiguous structures, such as the heart, the liver, and the kidneys, possess CT characteristics that are similar to those of the tumoral tissues. The parameters required for the segmentation procedure should reflect the characteristics of the entire tumor mass. For this reason, large values of standard deviation need to be used to characterize the wide range in the CT values of both diffuse and calcified regions. The heterogeneity and similarity, as described above, combine to impede the ability of the algorithm to halt the procedure effectively once the extent of the tumor has been reached:

the algorithm continues to aggregate falsely neighboring elements, resulting in leakage. In addition, the segmentation procedure was observed to be sensitive to seed selection. Seeds located in the intermediate-density tissues or calcified regions of the tumor were observed to produce different results in the work of Deglint et al. [11].

Attempts to segment the tumor mass directly result in severe leakage of the resulting region into contiguous anatomical structures [11, 12]. Furthermore, viable tumor, necrosis, fibrosis, and normal tissues are often intermixed within the tumor mass. Rather than attempt to separate these tissue types into distinct regions, Deglint et al. [11] proposed methods to delineate some of the normal structures expected in abdominal CT images, remove them from further consideration, and examine the remaining parts of the images for the tumor mass using fuzzy connectivity with manually provided seed pixels. In order to improve the result of segmentation, Deglint et al. [11] identified several potential sources of leakage in the body, and developed methods to remove them from further consideration. The methods include steps for detection and removal of peripheral artifacts, the skin, the peripheral fat, the peripheral muscle, and the spinal canal. Rangayyan et al. [42] proposed methods for the identification, segmentation, and removal of the spinal canal, which was also observed to cause difficulties in delimiting the tumor.

Vu et al. [12, 103] proposed an improved procedure for segmentation of the peripheral muscle, and implemented segmentation procedures using the technique of reconstruction. Rangayyan et al. [44, 54] proposed methods for identification of the diaphragm and subsequent removal of the thoracic cavity. Delineation of the diaphragm can be used to block leakage of the result of segmentation into the thoracic cavity; the diaphragm also provides an effective landmark for the identification of several abdominal and thoracic organs. The use of reconstruction with a region marker [12] led to better results in terms of the average true-positive rate (80.2%), but the average false-positive rate (1262.6%) remained high due to leakage into several abdominal tissues and organs [54]. In this context, the identification and removal of the lower pelvic region can prevent some of the leakage in the result of segmentation of the tumor and reduce the false-positive rate.

7.3   SEGMENTATION OF NEUROBLASTIC TUMORS

The image data (see Section 3.2 for details) were preprocessed by removing peripheral artifacts, the skin, the peripheral fat, the peripheral muscle, the spinal canal, the vertebral column, the thoracic cavity, and the regions below the upper pelvic surface, before performing the segmentation of the neuroblastic tumor. The procedure for segmentation of the tumor mass in neuroblastoma is as follows [12, 103, 59]:

1. A region within the tumor mass is manually selected on a CT slice. This region should be selected such that the different tissue types present in the tumor (that is, necrosis, calcification, and viable tumor) are all captured.

2. The mean, μ, and the standard deviation, σ, of the selected region are calculated.

3. The image is mapped to the fuzzy domain using a fuzzy membership function to obtain the mask for reconstruction. The fuzzy membership function, $m_A(r)$, given in Equation 2.23, is a mapping of the HU value of a pixel (or voxel) r in the CT image, $r \in D$, into the range $[0, 1]$. Fuzzy membership is used to represent the degree of similarity between an arbitrary pixel, r, and the object of interest, represented by its mean μ and standard deviation σ.

4. The whole selected region of the tumor is assigned to be the marker. Regardless of its size or shape, the marker must be contained within the mask; that is, if the mask is an image I and the marker is denoted by J, then $J \subseteq I$ must be satisfied, as described in Section 2.10.3.

5. Reconstruction is performed using the fuzzy-membership map as the mask and the selected region within the tumor as the marker. Although the fuzzy-membership map is a 3D volume, the marker could be a 2D region or a single seed voxel.

6. The result is thresholded to obtain a binary volume of the tumor. The threshold is based on the statistics calculated from the voxels within the user-selected region as

$$T = \mu_R + K\sigma_R, \tag{7.1}$$

where μ_R and σ_R are the mean and the standard deviation, respectively, calculated from the voxels that comprise the user-selected region applied to the fuzzy segmented region, and $K \geq 0$ is a constant. For the purpose of tumor segmentation $K = 0.5$ is used in this work.

7. The thresholded volume is morphologically closed to fill holes and gaps within the tumor, and to smooth the tumor volume.

The complete procedure for tumor segmentation is presented in the flowchart shown in Figure 7.1. It should be noted that in the present work only one marker region is provided on one CT slice, for each volumetric CT exam. The initial data provided in 2D are used to perform segmentation of the tumor in 3D. However, it is possible to provide multiple markers in several CT slices.

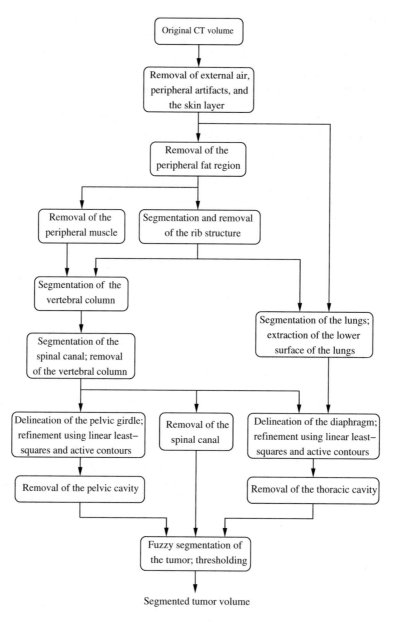

Figure 7.1: Flowchart describing the steps in the process of landmarking and segmentation of neuroblastic tumors. Reproduced with kind permission of Springer Science+Business media and from R M Rangayyan, S Banik, and G S Boag. "Landmarking and segmentation of computed tomographic images of pediatric patients with neuroblastoma." *International Journal of Computer Assisted Radiology and Surgery*, 2009. In press. © Springer.

7.4 ANALYSIS OF THE RESULTS

7.4.1 QUALITATIVE ANALYSIS

The procedure for segmentation of the neuroblastic tumor was applied to 10 CT exams of four patients as shown in Table 3.1. Results of the segmentation procedure, as above, are demonstrated in Figures 7.2, 7.3, and 7.4, for three selected CT exams (Exam 1C, Exam 1A, and Exam 2B). A region within the tumor was selected by the user, and the parameters were computed automatically as illustrated in Figures 7.2 (b), 7.3 (b), and 7.4 (b). After applying the reconstruction-based segmentation procedure, as shown in Figures 7.2 (c), 7.3 (c), and 7.4 (c), the fuzzy volumes were thresholded and morphologically closed using a 3D ellipsoidal structuring element with the dimensions $15 \times 15 \times 5$ voxels. For each exam being illustrated, the final result is shown in part (d) of the corresponding figure. The tumor mass in Figure 7.2 is highly calcified and nearly homogeneous, producing good results with the segmentation method. However, for the heterogeneous mass in Figure 7.3 (a), severe leakage into the liver, the spleen, the stomach, and the intestine occurred in the result obtained using the segmentation procedure. Due to the diffuse characteristics of the tumor (see Table 3.1), and the presence of low-density necrosis and medium-density viable tumor in the mass shown in Figure 7.4 (a), the results included other regions in the abdomen. The effects of prior removal of the vertebral column and the spinal canal are evident in the figures, as the results do not include any part of the vertebral column or the spinal canal. The 3D illustrations of the corresponding segmented tumor volumes are presented in Figures 7.5, 7.6, and 7.7.

The tumor mass shown in Figure 7.5 is highly calcified and nearly homogeneous, producing good results. A 3D view of the tumor mass as defined by the radiologist is shown in Figure 7.5 (a). The corresponding result of segmentation of the tumor, obtained with prior removal of peripheral artifacts, the skin, the peripheral fat, the peripheral muscle, and the spinal canal, and using the procedure described in Section 7.3, is shown in Figure 7.5 (b). The tumor volume is underestimated in Figure 7.5 (b); however, the result is comparable to the true tumor mass.

When the tumor is heterogeneous, as seen in Figure 7.3 and Figure 7.6, severe leakage into other tissues occurred; the liver, the spleen, the stomach, and the aorta have all been included in the final result. For diffuse tumors, such as that in Figure 7.7, the result of segmentation is poor due to leakage into the pelvic cavity and into the thoracic cavity. The mass as manually defined by the radiologist is shown in 3D in Figure 7.7 (a). Due to partial contrast uptake in some organs and contiguous structures, as can be seen in Figure 7.4 (a) and Figure 7.7 (b), the result of segmentation has leaked severely through the para-vertebral muscle into the thoracic cavity, and through colonic tissue into the pelvic cavity.

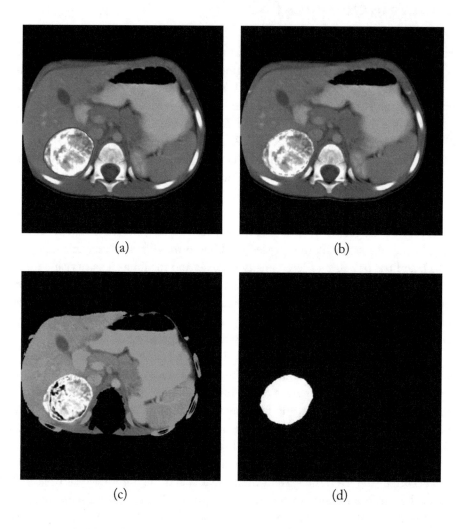

(a)   (b)

(c)   (d)

Figure 7.2: Exam 1C: (a) A CT slice of a two-year-old male patient after removing peripheral artifacts, with the manually drawn contour of the tumor by the radiologist shown in red line. (b) The same CT slice with the user-selected tumor region shown in dashed green line. (c) Result of reconstruction using the region marker shown in (b). (d) Result of thresholding and morphological closing of the result in (c). Note that the tumor is highly calcified and nearly homogeneous, leading to a good result. Reproduced with kind permission of Springer Science+Business media and from R M Rangayyan, S Banik, and G S Boag. "Landmarking and segmentation of computed tomographic images of pediatric patients with neuroblastoma." *International Journal of Computer Assisted Radiology and Surgery*, 2009. In press. © Springer. See Figure 7.5 for related 3D illustrations.

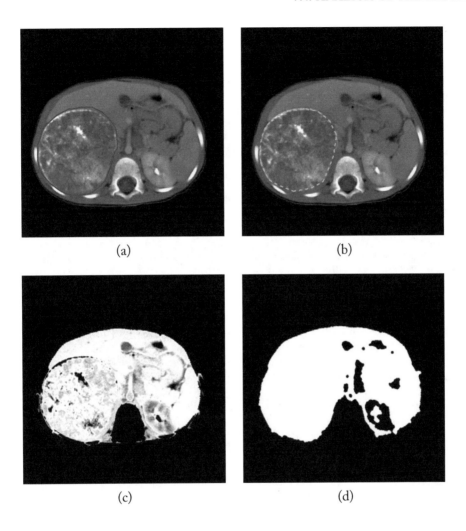

(a) (b)

(c) (d)

Figure 7.3: Exam 1A: (a) A CT slice of a two-year-old male patient after removing peripheral artifacts, with the manually drawn contour of the tumor by a radiologist shown in red line. (b) The same CT slice with the user-selected tumor region shown in dashed green line. (c) Result of reconstruction using the region marker shown in (b). (d) Result of thresholding and morphological closing of the result in (c). Due to the heterogeneity of the tumor, severe leakage has occurred into all other abdominal organs. The effect of removing the vertebral column prior to tumor segmentation is evident. Reproduced with kind permission of Springer Science+Business media and from R M Rangayyan, S Banik, and G S Boag. "Landmarking and segmentation of computed tomographic images of pediatric patients with neuroblastoma." *International Journal of Computer Assisted Radiology and Surgery*, 2009. In press. © Springer. See Figure 7.6 for related 3D illustrations.

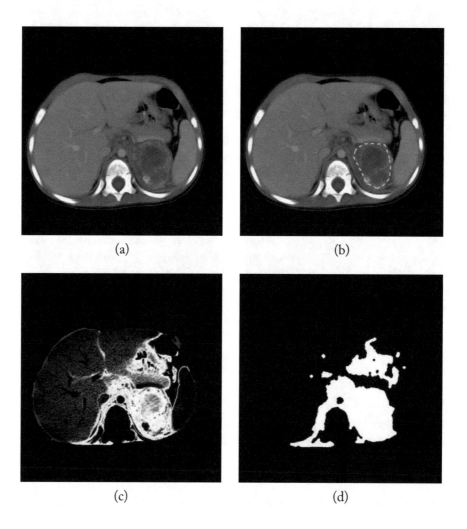

(a)   (b)

(c)   (d)

Figure 7.4: Exam 2B: (a) A CT slice of a two-year-old female patient after removing peripheral artifacts, with the manually drawn contour of the tumor by a radiologist shown in solid red line. (b) The same CT slice with the user-selected tumor region shown in dashed green line. (c) Result of reconstruction using the region marker shown in (b). (d) Result of thresholding and morphological closing of the result in (c). Due to the heterogeneity and diffuse nature of the tumor, leakage has occurred into other abdominal regions. See Figure 7.7 for related 3D illustrations.

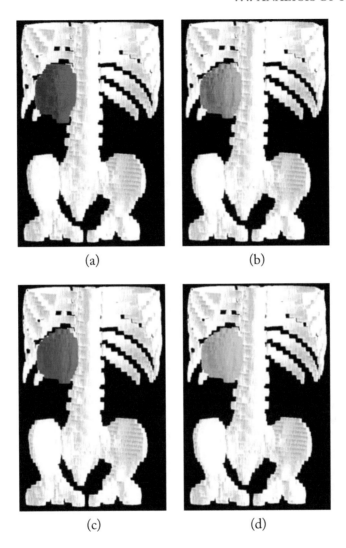

(a)   (b)

(c)   (d)

Figure 7.5: 3D rendition of the result of segmentation of the nearly homogeneous tumor in Exam 1C: (a) Tumor segmented manually by the radiologist, shown in red. (b) Result of segmentation of the tumor, shown in brown, with prior removal of peripheral artifacts, peripheral tissues, and the spinal canal. (c) Result of segmentation with prior removal of the thoracic cavity, shown in blue. (d) Result of segmentation with prior removal of the vertebral column and all regions below the upper pelvic surface, shown in green. Reproduced with kind permission of Springer Science+Business media and from R M Rangayyan, S Banik, and G S Boag. "Landmarking and segmentation of computed tomographic images of pediatric patients with neuroblastoma." *International Journal of Computer Assisted Radiology and Surgery*, 2009. In press. © Springer. See Figure 7.2 for related 2D illustrations.

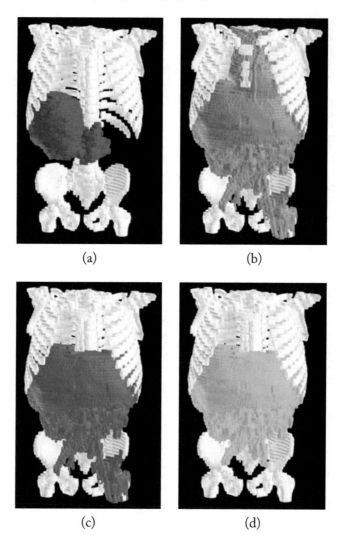

(a)   (b)

(c)   (d)

Figure 7.6:   3D rendition of the result of segmentation of the heterogeneous tumor in Exam 1A: (a) Tumor segmented manually by the radiologist, shown in red. (b) Result of segmentation of the tumor, shown in brown, with prior removal of peripheral artifacts, peripheral tissues, and the spinal canal. (c) Result of segmentation with prior removal of the thoracic cavity, shown in blue. (d) Result of segmentation with prior removal of the vertebral column and all regions below the upper pelvic surface, shown in green. Reproduced with kind permission of Springer Science+Business media and from R M Rangayyan, S Banik, and G S Boag. "Landmarking and segmentation of computed tomographic images of pediatric patients with neuroblastoma." *International Journal of Computer Assisted Radiology and Surgery*, 2009. In press. © Springer. See Figure 7.3 for related 2D illustrations.

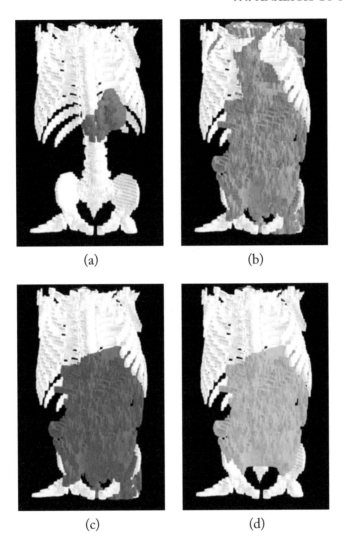

(a)   (b)

(c)   (d)

Figure 7.7:   3D rendition of the result of segmentation of the diffuse tumor in Exam 2B: (a) Tumor segmented manually by the radiologist, shown in red. (b) Result of segmentation of the tumor, shown in brown, with prior removal of peripheral artifacts, peripheral tissues, and the spinal canal. (c) Result of segmentation with prior removal of the thoracic cavity, shown in blue. (d) Result of segmentation with prior removal of the vertebral column and all regions below the upper pelvic surface, shown in green. See Figure 7.4 for related 2D illustrations. Figures in (b) and (d) are reproduced with permission from S Banik, R M Rangayyan, and G S Boag. "Delineation of the pelvic girdle in computed tomographic images". In *Proceedings of the 21st IEEE Canadian Conference on Electrical and Computer Engineering*, pages 179–182, Niagara Falls, Ontario, Canada, May 2008. © [2008] IEEE.

7.4.1.1 Use of the Spinal Canal and the Vertebral Column in Tumor Segmentation:

The spinal canal possesses CT characteristics similar to those of viable tumor, and has been identified to be a potential source of leakage in the process of tumor segmentation [11, 12]. If the result of segmentation includes the spinal canal, it introduces severe leakage by connecting other regions in the whole body in the CT exam. In this context, the segmented spinal canal (see Section 4.5) is removed along with the peripheral structures prior to the segmentation of the tumor.

The vertebral column is a bony structure and is not expected to be included in the result of segmentation of the neuroblastic tumor. However, due to the partial-volume effect and poor resolution, some of the vertebrae possess lower HU values than expected, and have been observed to be erroneously included in the result of segmentation. To prevent this error, the vertebral column, segmented by the method described in Section 4.4, is removed from each of the CT exam before performing the segmentation of the tumor.

The effects of prior removal of the vertebral column and the spinal canal are evident in Figures 7.2, 7.3, and 7.4 as the results do not include any part of the vertebral column or the spinal canal.

7.4.1.2 Use of the Diaphragm to Remove the Thoracic Region from Consideration:

If the tumor is heterogeneous or diffuse, significant leakage in the result of segmentation may occur into the thoracic cavity; as a result, the false-positive error rate becomes high in the result of segmentation. To prevent the leakage into the thoracic cavity, the diaphragm obtained by the procedure described in Section 5.3, is used in the present work as a separating boundary between the thoracic and the abdominal cavities; using the information related to the diaphragm, the thoracic cavity is removed from further consideration. As a result, the false-positive regions are reduced for the corresponding CT exams, as can be seen in Figures 7.6 (c) and 7.7 (c). The prior removal of the thoracic cavity resulted in significant reductions in the false-positive error rates without significantly affecting the true-positive rate.

If the tumor is contained within the abdomen and does not extend to the thoracic cavity, the procedure can assist in reducing the false-positive regions. However, if the tumor is nearly homogeneous and the result of the segmentation does not extend into the thoracic cavity, the procedure will have no effect, as seen in Figure 7.5 (c).

7.4.1.3 Use of the Pelvic Girdle to Reduce False Positives in the Results:

If the tumor is heterogeneous or diffuse, significant leakage in the result of segmentation may occur into the pelvic cavity, as can be seen in Figures 7.6 (b) and 7.7 (b); as a result, the false-positive error rate becomes high in the result of segmentation, because of inclusion of the lower pelvic organs and the intestines. To prevent the leakage into the lower pelvic region, the upper surface of the pelvic girdle, as obtained by the procedure described in Section 6.2, is used in the present work as a separating boundary between the abdominal and lower pelvic regions. The lower pelvic region is removed from further consideration using the upper surface of the pelvic girdle. As a result, the false-positive areas are reduced for the corresponding CT exams, as can be seen in Figures 7.6 (d) and 7.7 (d). The prior removal of the lower pelvic region resulted in significant reductions in the false-positive error rates without affecting the true-positive rate.

If the tumor is contained within the abdomen and does not extend to the pelvic cavity, the procedure can assist in reducing the false-positive regions. However, if the tumor is nearly homogeneous and the result of the segmentation does not extend into the pelvic cavity, the procedure will have no effect, as seen in Figure 7.5 (d).

7.4.2   QUANTITATIVE ANALYSIS

Quantitative analysis of the results of segmentation of the tumor using the measures of total error rate, false-positive error rate, and true-positive rate, as discussed in Section 3.3.2.3, are presented in Table 7.1 for the 10 CT exams processed. The average of each measure is calculated by taking the mean of the corresponding measure over the 10 CT exams. The percentage change in each error measure is calculated as follows:

$$\%Change = \frac{\varepsilon_{n+1} - \varepsilon_n}{\varepsilon_n} \times 100\%, \tag{7.2}$$

where ε_n is the percentage error measure after the application of the previous step and ε_{n+1} is the percentage error measure after the application of each additional step, such as the removal of the thoracic cavity (step-II) and the removal of the vertebral column and the pelvic cavity (step-III). The average of the percentage change in each error measure is calculated by taking the mean of the corresponding percentage change over the 10 CT exams.

The effect of removal of the thoracic cavity in the 10 CT exams is illustrated in Part (II) of Table 7.1. The table shows that the steps have improved the segmentation results by reducing the false-positive rate. The method prevents leakage of the result of segmentation of the tumor into the thoracic region, and restricts the tumor segmentation process to the abdominal cavity. In nine of the 10 CT exams, the methods described in the present work resulted in a reduction in the false-positive rate. In the remaining one exam, there was no improvement in the result of segmentation because the tumor mass is well-defined, and the result did not extend beyond the abdominal cavity; therefore, the removal of the thoracic region did not affect the result. In cases where there was significant leakage into the thoracic region, delineating the diaphragm and removing the thorax from consideration resulted in significant reductions in the false-positive rates. The true-positive rates remained unchanged in nine CT exams.

The results obtained in the present work are slightly better than those reported by Rangayyan et al. [54] and Vu et al. [12] with the same dataset. Due to the modification of the segmentation process for the peripheral muscle, the lungs, and the diaphragm, and the additional step of prior removal of the vertebral column, the results obtained are improved in terms of the false-positive error rate and total error rate. The true-positive rate remained unchanged. Note that, in Exam 3B, there is a significant reduction in error after removal of the thoracic cavity. Because of the improved segmentation of the peripheral muscle and the modified process of segmentation of the diaphragm, the leakage into the thoracic cavity is reduced. The prior removal of the vertebral column and the rib

Table 7.1: The effect of removal of the interfering regions, the thoracic cavity, the vertebral column, and the regions inferior to the upper pelvic surface. The total error (ε_T), the false-positive error rate (ε_{FP}), and the true-positive rate (ε_{TP}) are all in percentages with respect to the corresponding tumor volume segmented by the radiologist. (I) Initial result of segmentation of the tumor with prior removal of the peripheral artifacts, the skin, the peripheral fat, the peripheral muscle, and the spinal canal. (II) Result of tumor segmentation with prior removal of the diaphragm and the thoracic cavity from further consideration. (III) Result of segmentation with prior removal of the vertebral column and the region below the upper pelvic surface. The true-positive rate (ε_{TP}) is not affected by the method described (III). Reproduced with kind permission of Springer Science+Business media and from R M Rangayyan, S Banik, and G S Boag. "Landmarking and segmentation of computed tomographic images of pediatric patients with neuroblastoma." *International Journal of Computer Assisted Radiology and Surgery*, 2009. In press. © Springer.

Exam	Initial result of segmentation (I)			Result of segmentation with prior removal of the diaphragm (II)					Result of segmentation with prior removal of the vertebral column and the region inferior to the upper pelvic surface (III)			
	ε_T	ε_{FP}	ε_{TP}	ε_T	ε_{FP}	ε_{TP}	Change % in ε_T	Change % in ε_{FP}	ε_T	ε_{FP}	Change % in ε_T	Change % in ε_{FP}
1A	244.9	265.7	79.2	200.6	221.4	79.2	-18.1	-16.7	180.7	201.7	-9.9	-8.9
1B	861.4	868.1	93.4	681.7	688.4	93.4	-20.9	-20.7	560.6	567.2	-17.8	-17.6
1C	-15.6	0.8	83.5	-15.6	0.8	83.5	0.0	0.0	-15.6	0.8	0.0	0.0
2A	115.3	127.7	87.5	95.3	109.4	85.9	-17.3	-14.3	59.6	73.7	-37.5	-32.6
2B	1376.9	1382.1	94.8	1118.7	1123.9	94.8	-18.8	-18.7	898.4	903.6	-19.7	-19.6
3A	1399.9	1416.9	83.0	1385.2	1402.2	83.0	-1.1	-1.0	1385.0	1402.0	0.0	0.0
3B	7552.0	7610.4	41.6	7066.2	7124.5	41.6	-6.4	-6.4	7066.2	7124.5	0.0	0.0
4A	-8.4	18.6	73.0	-9.2	18.0	73.0	9.5	-3.2	-10.4	17.0	13.0	-5.6
4B	683.4	690.1	93.3	674.1	680.9	93.3	-1.4	-1.3	551.6	558.3	-18.2	-18.0
4D	678.1	684.3	93.8	672.1	678.3	93.8	-0.9	-0.9	670.7	676.9	-0.2	-0.2
Average	1288.8	1306.5	82.3	1186.9	1204.8	82.1	-7.5	-8.3	1134.7	1152.6	-9.0	-10.3

structure in the procedure of segmentation of the diaphragm is also indirectly responsible for this improvement.

The effect of removal of the vertebral column and the regions inferior to the upper pelvic surface for the 10 CT exams processed is illustrated in Part (III) of Table 7.1. The table shows that the method has improved the segmentation results by reducing the false-positive error rate. The method prevents leakage of the result of segmentation of the tumor into the lower region of the pelvis, and restricts the tumor segmentation process to the abdominal cavity. In seven of the 10 CT exams, the methods described in the present work resulted in a reduction in the false-positive rate. In the remaining three exams, there was no improvement in the result of segmentation because the tumor masses are well-defined, and the result did not extend below the abdominal cavity; therefore, the removal of the lower pelvic region did not affect the result. In cases where there was significant leakage into the lower abdomen, delineating the pelvic girdle and removing the region below the upper pelvic surface resulted in significant reductions in the false-positive error rates. The true-positive rates remained unchanged in all cases. The removal of the vertebral column assisted in reducing the false-positive error rates in four CT exams; especially, in Exam 1B, the false-positive rate was reduced by 10.5%. On the average, the removal of the vertebral column resulted in 3.1% reduction in the false-positive error rate over the 10 CT exams processed. Overall, the methods for removal of the vertebral column, the thoracic cavity, and the pelvic cavity resulted in the reduction of the false-positive error rates by 22.4%, on the average, over the 10 CT exams processed.

7.5   REMARKS

The success of segmentation of the neuroblastic tumor depends largely on the location of the tumor (in the abdomen) and the tissue composition of the tumor. The methods for delineation of the peripheral structures, the spinal canal, the vertebral column, the diaphragm, and the pelvic girdle have assisted in reducing the false-positive error rate in the segmentation of abdominal tumors due to neuroblastoma, and have provided promising results.

Due to heterogeneity of the tumor, in some cases, the result of segmentation includes leakage into other abdominal organs. When the tumor is located in the abdomen, and leakage into the thoracic or the lower pelvic region is significant, the methods described above are able to prevent leakage into the thorax and the lower pelvic region in the region growing method. However, the methods will not be applicable if the neuroblastic tumor originates in or extends to the thoracic region (such as in Patient 11) or extends to the lower pelvic region (such as in Patient 8); the rates of occurrence of such tumors are low.

CHAPTER 8

Concluding Remarks

The segmentation and delineation of different organs in CT images are difficult tasks; especially in the case of CT exams of children affected by cancer, due to the unavailability of standard atlases to facilitate the atlas-based segmentation procedures, the segmentation of different organs is challenging. The application of landmarking procedures with prior knowledge appears to be effective, as demonstrated in the present work. Several landmarks can be used in the process of localization, identification, and segmentation of thoracic, abdominal, and pelvic organs in pediatric CT images. The results obtained by the presented methods for segmentation of peripheral artifacts, the skin, the peripheral fat, the peripheral muscle, the ribs, the vertebral column, the spinal canal, the lungs, the diaphragm, and the pelvic girdle have been assessed in a subjective manner as well as an objective manner. All of the procedures discussed for segmentation of organs have provided good results.

The segmentation of the tumor mass in neuroblastoma is a challenging problem. The commonly heterogeneous mass creates many problems for different segmentation algorithms in the form of leakage. In the present work, several procedures were developed, modified, and applied to the problem of segmentation with the aim of reducing the error rates and improving tumor definition. To improve the results of segmentation, new procedures to remove the vertebral column and the lower pelvic region were described. The methods resulted in reduced false-positive rates and assisted in improving the definition of the segmented tumor volume.

Landmarking of CT images is challenging as well as interesting work. Several other landmarks in the abdominal and thoracic region, such as the kidneys, the liver, the heart, and the intestines could be incorporated in the future. The pelvic surface may be used to localize lower pelvic organs, such as the prostate. After successful segmentation of several other landmarks, it might be possible to build an abdominal probabilistic atlas for pediatric CT exams. The methods may also be extended to other imaging modalities.

For the purpose of segmentation of neuroblastic tumors, additional work is required to reduce the leakage of the result of segmentation into contiguous tissues and anatomical structures. The final result of segmentation is dependent on the threshold applied to the result of reconstruction. Investigation into the derivation of an optimal or appropriate threshold for application to the result of reconstruction could help to reduce the error in the final results. The use of a receiver operating characteristics (ROC) curve [129] could also help to quantify the behavior of the segmentation procedure in response to variations in the threshold and the other parameters in the procedures.

Future paths related to this work could include a study on the use of simultaneous and competitive region growing methods [21, 22, 36] to segment other abdominal organs as well as the primary tumors in neuroblastoma. Once segmentation of the tumor mass is achieved, a Gaussian mixture model [17] could be applied to extract quantitative information corresponding to the tumor's tissue composition.

Typically, in response to therapy, the tumor mass shrinks in size and the tissue composition changes. A shift in the pattern of the segmented tumor's histogram from predominantly intermediate-density tumoral tissues to predominantly low-attenuation necrotic tissue, and ultimately to predominantly high-attenuation calcified tissue represents a good response to therapy. In contrast, the absence of such a progression may be an indicator of residual viable tumor or justification for modified therapy [17]. Quantifying such information could aid the radiologist in assessing a patient's response to therapy, and may help in guiding subsequent therapy. The methods can also be extended to other imaging modalities for neuroblastoma, as well as to studies of other tumors.

Bibliography

[1] H Barrett and W Swindell. *Radiological Imaging: The Theory of Image Formation, Detection and Processing*, volume 1-2. Academic Press, New York, NY, 1981.

[2] A P Dhawan. *Medical Image Analysis*. IEEE Press, Piscataway, NJ, 2003.

[3] K Doi. Diagnostic imaging over the last 50 years: research and developement in medical imaging science and technology. *Physics in Medicine and Biology*, 51:R5–R27, June 2006. DOI: 10.1088/0031-9155/51/13/R02

[4] K Doi. Computer-aided diagnosis in medical imaging: Historical review, current status and future potential. *Computerized Medical Imaging and Graphics*, 31(4-5):198–211, June-July 2007. DOI: 10.1016/j.compmedimag.2007.02.002

[5] R M Rangayyan. *Biomedical Image Analysis*. CRC Press LLC, Boca Raton, FL, 2005.

[6] L V Petrus, T R Hall, M I Boechat, S J Westra, J G Curran, R J Steckel, and H Kangarloo. The pediatric patient with suspected adrenal neoplasm: Which radiological test to use? *Medical and Pediatric Oncology*, 20:53–57, 1992. DOI: 10.1002/mpo.2950200111

[7] C R Staalman and C A Hoefnagel. Imaging of neuroblastoma and metastasis. In G M Brodeur, T Sawada, Y Tsuchida, and P A Voûte, editors, *Neuroblastoma*, pages 303–340. Elsevier, Amsterdam, The Netherlands, 2000.

[8] K D Hopper, K Singapuri, and A Finkel. Body CT and oncologic imaging. *Radiology*, 215(1):27–40, 2000.

[9] J R Sty, R G Wells, R J Starshak, and D C Gregg. *Diagnostic Imaging of Infants and Children*, volume I. Aspen Publishers, Inc., Gaithersburg, MD, 1992.

[10] G M Brodeur, J Pritchard, F Berthold, N L T Carlsen, V Castel, R P Castleberry, B De-Bernardi, A E Evans, M Favrot, F Hedborg, M Kaneko, J Kemshead, F Lampert, R E J Lee, A T Look, A D J Pearson, T Philip, B Roald, T Sawada, R C Seeger, Y Tsuchida, and P A Voûte. Revisions of the international criteria for neuroblastoma diagnosis, staging, and response to treatment. *Journal of Clinical Oncology*, 11(8):1466–1477, August 1993.

[11] H J Deglint, R M Rangayyan, F J Ayres, G S Boag, and M K Zuffo. Three-dimensional segmentation of the tumor in computed tomographic images of neuroblastoma. *Journal of Digital Imaging*, 20(1):72–87, 2007. DOI: 10.1007/10278-006-0769-3

[12] R H Vu, R M Rangayyan, H J Deglint, and G S Boag. Segmentation and analysis of neuroblastoma. *Journal of the Franklin Institute*, 344(3-4):257–284, May-July 2007. DOI: 10.1016/j.jfranklin.2006.11.002

[13] B H Kushner. Neuroblastoma: A disease requiring a multitude of imaging studies. *Journal of Nuclear Medicine*, 45(7):101–105, July 2004.

[14] S J Abramson. Adrenal neoplasms in children. *Radiologic Clinics of North America*, 35(6):1415–1453, 1997.

[15] D J Goodenough. Tomographic imaging. In J Beutel, H L Kundel, and R L Van Metter, editors, *Handbook of Medical Imaging, Volume 1: Physics and Psychophysics*, pages 511–558. SPIE Press, Bellingham, WA, 2000.

[16] G D Fullerton. Fundamentals of CT tissue characterization. In G D Fullerton and J A Zagzebski, editors, *Medical Physics of CT and Ultrasound: Tissue Imaging and Characterization*, pages 125–162. American Association of Physicists in Medicine, New York, NY, 1980.

[17] F J Ayres, M K Zuffo, R M Rangayyan, G S Boag, V Odone Filho, and M Valente. Estimation of the tissue composition of the tumor mass in neuroblastoma using segmented CT images. *Medical and Biological Engineering and Computing*, 42:366–377, 2004. DOI: 10.1007/BF02344713

[18] V C Mategrano, J Petasnick, J Clark, A C Bin, and R Weinstein. Attenuation values in computed tomography of the abdomen. *Radiology*, 125:135–140, October 1977.

[19] H Park, P H Bland, and C R Meyer. Construction of an abdominal probabilistic atlas and its application in segmentation. *IEEE Transactions on Medical Imaging*, 22(4):483–492, April 2003. DOI: 10.1109/TMI.2003.809139

[20] H Kobatake. Future CAD in multi-dimensional medical images - project on multi-organ, multi-disease CAD system. *Computerized Medical Imaging and Graphics*, 31(4-5):258–266, 2006.

[21] S Hugueny and M Rousson. Hierarchical detection of multiple organs using boosted features. In W G Kropatsch, M Kampel, and A Hanbury, editors, *Computer Analysis of Images and Patterns*, volume 4673 of *Lecture Notes in Computer Science*, pages 317–325. Springer, Berlin, Germany, 2007. DOI: 10.1007/978-3-540-74272-2

[22] A Shimizu, R Ohno, T Ikegami, H Kobatake, S Nawano, and D Smutek. Multi-organ segmentation in three dimensional abdominal CT images. In *Proceedings of CARS 20th International Congress and Exhibition: Computer Assisted Radiology and Surgery*, volume 1, pages 76–78, Osaka, Japan, 2006.

[23] C C Lee, P C Chung, and H M Tsai.   Identifying multiple abdominal organs from CT image series using a multimodule contextual neural network and spatial fuzzy rules. *IEEE Transactions on Information Technology in Biomedicine*, 7(3):208–217, September 2003. DOI: 10.1109/TITB.2003.813795

[24] H Fujimoto, L Gu, and T Kaneko. Recognition of abdominal organs using 3D mathematical morphology. *Systems and Computers in Japan*, 33(8):75–83, 2002. DOI: 10.1002/scj.1148

[25] M Kobashi and L G Shapiro. Knowledge-based organ identification from CT images. *Pattern Recognition*, 28(4):475–491, 1995. DOI: 10.1016/0031-3203(94)00124-5

[26] A Hill, C J Taylor, and A D Brett. A framework for automatic landmark identification using a new method of nonrigid correspondence. *IEEE Transactions on Pattern Analysis and Machine Intelligence*, 22(3):241–251, March 2000. DOI: 10.1109/34.841756

[27] M S Brown and M F McNitt-Gray. Medical image interpretation. In M Sonka and J M Fitzpatrick, editors, *Handbook of Medical Imaging, Volume 2: Medical Image Processing and Analysis*, pages 399–445. SPIE Press, Bellingham, WA, 2000.

[28] N Archip, P J Erard, M E Petersen, J M Haefliger, and J F Germond. A knowledge-based approach to automatic detection of the spinal cord in CT images. *IEEE Transactions on Medical Imaging*, 21(12):1504–1516, December 2002. DOI: 10.1109/TMI.2002.806578

[29] N Karssemeijer, L J T O van Erning, and E G J Eijkman. Recognition of organs in CT image sequences: A model guided approach. *Computers and Biomedical Research*, 21(5):434–448, October 1988. DOI: 10.1016/0010-4809(88)90003-1

[30] S M Qatarneh, J Crafoord, E L Kramer, G Q Maguire Jr, A Brahme, M E Noz, and S Hyödynmaa. A whole body atlas for segmentation and delineation of organs for radiation therapy planning. *Nuclear Instruments and Methods in Physics Research*, A471(1):160–164, 2001. DOI: 10.1016/S0168-9002(01)00984-6

[31] S M Qatarneh, M E Noz, S Hyödynmaa, G Q Maguire, E L Kramer, and J Crafoord. Evaluation of a segmentation procedure to delineate organs for use in construction of a radiation therapy planning atlas. *International Journal of Medical Informatics*, 69(1):39–55, 2003. DOI: 10.1016/S1386-5056(02)00079-5

[32] T Rohlfing, D B Russakoff, and C R Maurer. Performance-based classifier combination in atlas-based image segmentation using expectation-maximization parameter estimation. *IEEE Transactions on Medical Imaging*, 23(8):983–994, August 2004. DOI: 10.1109/TMI.2004.830803

[33] J Ehrhardt, H Handels, T Malina, B Strathmann, W Plötz, and S J Pöppl. Atlas-based segmentation of bone structures to support the virtual planning of hip operations. *International Journal of Medical Informatics*, 64(2):439–447, December 2001. DOI: 10.1016/S1386-5056(01)00212-X

[34] J Ehrhardt, H Handels, W Plötz, and S J Pöppl. Atlas-based recognition of the anatomical structures and landmarks and the automatic computation of the orthopedic parameters. *Methods of Information in Medicine*, 43(4):391–397, 2004.

[35] X Zhou, N Kamiya, T Hara, H Fujita, H Chen, R Yokoyama, and H Hoshi. Automated segmentation and recognition of abdominal wall muscles in x-ray torso CT images and its application in abdominal CAD. In *Proceedings of CARS 21st International Congress and Exhibition: Computer Assisted Radiology and Surgery*, volume 2, pages 388–390, 2007.

[36] A Shimizu, H Sakurai, H Kobatake, S Nawano, and D Smutek. Improvement of a multi-organ extraction algorithm in an abdominal CAD system based on features in neighbouring regions. In *Proceedings of CARS 21st International Congress and Exhibition: Computer Assisted Radiology and Surgery*, volume 2(1), pages 386–388, Berlin, Germany, June 2007.

[37] J D Furst, R Susomboom, and D S Raicu. Single organ segmentation filters for multiple organ segmentation. In *Proceedings of the 28th Annual International Conference of the IEEE Engineering in Medicine and Biology Society*, pages 3033 – 3036, Lyon, France, August 2006.

[38] J Yao, S D O'Connor, and R M Summers. Automated spinal column extraction and partitioning. In *Proceedings of the 3rd IEEE International Symposium on Biomedical Imaging: Nano to Macro*, pages 390–393, Arlington, VA, April 2006. DOI: 10.1109/ISBI.2006.1624935

[39] H Wang, J Bai, and Y Zhang. A relative thoracic cage coordinate system for localizing the thoracic organs in chest CT volume data. In *Proceedings of the 27th Annual International Conference of the IEEE Engineering in Medicine and Biology Society*, pages 3257–3260, Shanghai, China, September 2005. DOI: 10.1109/IEMBS.2005.1617171

[40] J Staal, B V Ginneken, and M A Viergever. Automatic rib segmentation in CT data. In M Sonka, I A Kakadiaris, and J Kybic, editors, *Computer Vision and Mathematical Methods in Medical and Biomedical Image Analysis*, volume 3117/2004, pages 193–204. Springer, Berlin, Germany, 2004.

[41] G Karangelis and S Zimeras. 3D segmentation method of the spinal cord applied on CT data. *Computer Graphics Topics*, 14(1/2002):28–29, 2002.

[42] R M Rangayyan, H J Deglint, and G S Boag. Method for the automatic detection and segmentation of the spinal canal in computed tomographic images. *Journal of Electronic Imaging*, 15(3):033007–1:9, 2006. DOI: 10.1117/1.2234770

[43] M Hahn and T Beth. Balloon based vertebra separation in CT images. In *Proceedings of the 17th IEEE Symposium on Computer-Based Medical Systems*, pages 310–315, Los Alamitos, CA, June 2004. DOI: 10.1109/CBMS.2004.1311733

[44] R M Rangayyan, R H Vu, and G S Boag. Automatic delineation of the diaphragm in computed tomographic images. *Journal of Digital Imaging*, 21(1):S134–S147, November 2008. DOI: 10.1007/s10278-007-9091-y

[45] O Camara, O Colliot, and I Bloch. Computational modeling of thoracic and abdominal anatomy using spatial relationships for image segmentation. *Real-Time Imaging*, 10(4):263–273, August 2004. DOI: 10.1016/j.rti.2004.05.005

[46] D Y Kim and J W Park. Computer-aided detection of kidney tumor on abdominal computed tomography scans. *Acta Radiologica*, 45(7):791–795, 2004. DOI: 10.1080/02841850410001312

[47] M G Linguraru, J Yao, R Gautam, J Peterson, Z Li, W M Linehan, and R M Summers. Renal tumor quantification and classification in contrast-enhanced abdominal CT. *Pattern Recognition*, 2009. In press. DOI: 10.1016/j.patcog.2008.09.018

[48] M R Kaus, S K Warfield, A Nabavi, P M Black, F A Jolesz, and R Kikinis. Automated segmentation of MR images of brain tumors. *Radiology*, 218(2):586–591, February 2001.

[49] L Soler, H Delingette, G Malandain, J Montagnat, N Ayache, C Koehl, O Dourthe, B Malassagne, M Smith, D Mutter, and J Marescaux. Fully automatic anatomical, pathological, and functional segmentation from CT scans for hepatic surgery. *Computer Aided Surgery*, 6(3):131–142, October 2001. DOI: 10.1002/igs.1016

[50] F Alexander. Neuroblastoma. *Urologic Clinics of North America*, 27(3):383–392, August 2000. DOI: 10.1016/S0094-0143(05)70087-2

[51] M Schwab, F Westermann, B Hero, and F Berthold. Neuroblastoma: biology and molecular and chromosomal pathology. *The Lancet Oncology*, 4:472–80, 2003. DOI: 10.1016/S1470-2045(03)01166-5

[52] R P Castleberry. Neuroblastoma. *European Journal of Cancer*, 33(9):1430–1438, 1997. DOI: 10.1016/S0959-8049(97)00308-0

[53] G M Brodeur and J M Maris. Neuroblastoma. In P A Pizzo and D G Poplack, editors, *Principles and Practice of Pediatric Oncology*, pages 895–937. Lippincott Williams and Wilkins, Philadelphia, PA, 2002.

[54] R M Rangayyan, R H Vu, and G S Boag. Delineation of the diaphragm in CT images to improve segmentation of the tumor mass in neuroblastoma. In *Proceedings of CARS 20th*

International Congress and Exhibition: Computer Assisted Radiology, pages 78–80, Osaka, Japan, 2006.

[55] S Banik, R M Rangayyan, and G S Boag. Delineation of the pelvic girdle in computed tomographic images. In *Proceedings of the 21st IEEE Canadian Conference on Electrical and Computer Engineering*, pages 179–182, Niagara Falls, Ontario, Canada, May 2008. DOI: 10.1109/CCECE.2008.4564519

[56] S Banik, R M Rangayyan, and G S Boag. Landmarking of computed tomographic images to assist in segmentation of abdominal tumors caused by neuroblastoma. In *Proceedings of the 30th Annual International Conference of the IEEE Engineering in Medicine and Biology Society*, pages 3126–3129, Vancouver, BC, Canada, August 2008. DOI: 10.1109/IEMBS.2008.4649866

[57] S Banik, R M Rangayyan, G S Boag, and R H Vu. Segmentation of the pelvic girdle in pediatric computed tomographic images. *Journal of Electronic Imaging*, June 2008. Submitted.

[58] S Banik. Three-dimensional image processing techniques to perform landmarking and segmentation of computed tomographic images. Master's thesis, University of Calgary, Calgary, Alberta, Canada, April 2008.

[59] R M Rangayyan, S Banik, and G S Boag. Landmarking and segmentation of computed tomographic images of pediatric patients with neuroblastoma. *International Journal of Computer Assisted Radiology and Surgery*, 2009. In press.

[60] R M Rangayyan, S Banik, and G S Boag. Automatic segmentation of the ribs and the vertebral column in computed tomographic images of pediatric patients. In *Proceedings of CARS 22nd International Congress and Exhibition: Computer Assisted Radiology and Surgery*, volume 3(1), pages S42–S44, Barcelona, Spain, June 2008.

[61] S Banik, R M Rangayyan, and G S Boag. Automatic segmentation of the ribs, the vertebral column, and the spinal canal in pediatric computed tomographic images. *Journal of Digital Imaging*, 2009. In press.

[62] R C Gonzalez and R E Woods. *Digital Image Processing*. Prentice Hall, Upper Saddle River, NJ, second edition, 2002.

[63] J L Prince and J M Links. *Medical Imaging Signals and Systems*. Pearson Prentice Hall, Upper Saddle River, NJ, 2006.

[64] T Acharya and A K Ray. *Image Processing Principles and Applications*. John Wiley & Sons, Inc., Hoboken, NJ, 2005.

[65] E R Dougherty. *An Introduction to Morphological Image Processing*. SPIE Press, Bellingham, WA, 1992.

[66] R O Duda, P E Hart, and D G Stork. *Pattern Classification*. John Wiley and Sons, Inc., New York, NY, 2nd edition, 2001.

[67] J G Proakis and D G Manolakis. *Digital Signal Processing: Principles, Algorithms, and Applications*. Prentice Hall, Upper Saddle River, NJ, third edition, 1996.

[68] P K Sahoo, S Soltani, A K C Wong, and Y C Chen. A survey of thresholding techniques. *Computer Vision, Graphics, and Image Processing*, 41:233–260, 1988. DOI: 10.1016/0734-189X(88)90022-9

[69] B M Dawant and A P Zijdenbos. Image segmentation. In M Sonka and J M Fitzpatrick, editors, *Handbook of Medical Imaging, Volume 2: Medical Image Processing and Analysis*, pages 71–127. SPIE Press, Bellingham, WA, 2000.

[70] N Otsu. A threshold selection method from gray level histograms. *IEEE Transactions on Systems, Man and Cybernetics*, 9:62–66, 1979.

[71] T Pun. Entropic thresholding: a new approach. *Computer Graphics and Image Processing*, 16(3):210–239, 1980. DOI: 10.1016/0146-664X(81)90038-1

[72] J N Kapur, P K Sahoo, and A K C Wong. A new method for gray level picture thresholding using the entropy of histogram. *Computer Graphics, Vision, and Image Processing*, 29(3):273–285, 1985. DOI: 10.1016/0734-189X(85)90125-2

[73] A K C Wong and P K Sahoo. A gray level threshold selection method based on maximum entropy principle. *IEEE Transactions on Systems, Man, and Cybernetics*, 19(4):866–871, 1989. DOI: 10.1109/21.35351

[74] L Shen. *Region-based Adaptive Image Processing Techniques for Mammography*. PhD thesis, Department of Electrical and Computer Engineering, University of Calgary, Calgary, Alberta, Canada, August 1998.

[75] J Canny. A computational approach to edge detection. *IEEE Transactions on Pattern Analysis and Machine Intelligence*, 8(6):679–698, November 1986. DOI: 10.1109/TPAMI.1986.4767851

[76] C Xu, D L Pham, and J L Prince. Image segmentation using deformable models. In M Sonka and J M Fitzpatrick, editors, *Handbook of Medical Imaging, Volume 2: Medical Image Processing and Analysis*, pages 129–174. SPIE Press, Bellingham, WA, 2000.

[77] T McInerney and D Terzopoulos. Deformable models in medical image analysis: A survey. *Medical Image Analysis*, 1(2):91–108, 1996. DOI: 10.1016/S1361-8415(96)80007-7

[78] M Kass, A Witkin, and D Terzopoulos. Snakes: active contour models. *International Journal of Computer Vision*, 1(4):321–331, 1988. DOI: 10.1007/BF00133570

[79] C Xu and J L Prince. Snakes, shapes, and gradient vector flow. *IEEE Transactions on Image Processing*, 7(3):359–369, 1998.

[80] P V C Hough. A method and means for recognizing complex patterns. US Patent 3,069,654, December 18, 1962.

[81] R O Duda and P E Hart. Use of the Hough transform to detect lines and curves in pictures. *Communications of the ACM*, 15(1):11–15, January 1972. DOI: 10.1145/361237.361242

[82] C B Barber, D P Dobkin, and H Huhdanpaa. The quickhull algorithm for convex hulls. *ACM Transactions on Mathematical Software*, 22(4):469–483, 1996. DOI: 10.1145/235815.235821

[83] L A Zadeh. Fuzzy sets. *Information and Control*, 8:338–353, 1965. DOI: 10.1016/S0019-9958(65)90241-X

[84] J C Bezdek. *Fuzzy Models for Pattern Recognition: Methods that search for structures in data.* IEEE Press, New York, NY, 1992.

[85] J K Udupa and S Samarasekera. Fuzzy connectedness and object definition: Theory, algorithms, and applications in image segmentation. *Graphical Models and Image Processing*, 58(3):246–261, 1996. DOI: 10.1006/gmip.1996.0021

[86] A Rosenfeld. The fuzzy geometry of image subsets. *Pattern Recognition Letters*, 2(5):311–317, 1984. DOI: 10.1016/0167-8655(84)90018-7

[87] C R Giardina and E R Dougherty. *Morphological Methods in Image and Signal Processing.* Prentice Hall, Englewood Cliffs, NJ, 1988.

[88] J Goutsias and S Batman. Morphological methods for biomedical image analysis. In M Sonka and J M Fitzpatrick, editors, *Handbook of Medical Imaging, Volume 2: Medical Image Processing and Analysis*, pages 175–272. SPIE Press, Bellingham, WA, 2000.

[89] E R Dougherty and R A Lotufo. *Hands-on Morphological Image Processing*, volume TT59 of *Tutorial Texts in Optical Engineering*. SPIE Press, Bellingham, WA, 2003.

[90] H Blum. A transformation for extracting new descriptors of shape. In W Wathen-Dunn, editor, *Models for the Perception of Speech and Visual Form*. MIT Press, Cambridge, MA, 1967.

[91] H Blum. Biological shape and visual science (Part I). *Journal of Theoretical Biology*, 38:205–287, 1973. DOI: 10.1016/0022-5193(73)90175-6

[92] L Vincent. Efficient computation of various types of skeletons. In *Proceedings of the SPIE Conference on Medical Imaging V*, volume 1445, pages 297–311, San Jose, CA, 1991.

[93] J Serra. *Image Analysis and Mathematical Morphology.* Academic Press, New York, NY, 1983.

[94] L Vincent. Morphological grayscale reconstruction in image analysis: Applications and efficient algorithms. *IEEE Transactions on Image Processing*, 2(2):176–201, 1993. DOI: 10.1109/83.217222

[95] I Bloch. Fuzzy connectivity and mathematical morphology. *Pattern Recognition Letters*, 14:483–488, 1993. DOI: 10.1016/0167-8655(93)90028-C

[96] AVS: Advanced Visual Systems. http://www.avs.com. Online.

[97] ImageJ: Image processing and analysis in Java. http://rsb.info.nih.gov/ij. Online.

[98] G Rote. Computing the minimum Hausdorff distance between two point sets on a line under translation. *Information Processing Letters*, 38:123–127, 1991. DOI: 10.1016/0020-0190(91)90233-8

[99] J Xu, O Chutatape, and P Chew. Automated optic disk boundary detection by modified active contour model. *IEEE Transactions on Biomedical Engineering*, 54(3):473–482, 2007. DOI: 10.1109/TBME.2006.888831

[100] D P Huttenlocher, G A Klanderman, and W J Rucklidge. Comparing images using the Hausdorff distance. *IEEE Transactions on Pattern Analysis and Machine Intelligence*, 15(9):850–853, 1993. DOI: 10.1109/34.232073

[101] P K Saha, B B Chaudhuri, and D D Majumder. A new shape preserving parallel thinning algorithm for 3D digital images. *Pattern Recognition*, 30(12):1939–1955, 1997. DOI: 10.1016/S0031-3203(97)00016-2

[102] C C Lee and P C Chung. Recognizing abdominal organs in CT images using contextual neural network and fuzzy rules. In *Proceedings of the 22nd Annual International Conference of the IEEE Engineering in Medicine and Biology Society*, pages 1745–1748, Chicago, IL, July 2000.

[103] R H Vu. Strategies for three-dimensional segmentation of the primary tumor mass in computed tomographic images of neuroblastoma. Master's thesis, Schulich School of Engineering, University of Calgary, Calgary, Alberta, Canada, July 2006.

[104] T Mäkelä, P Clarysse, O Sipilä, N Pauna, Q C Pham, T Katila, and I E Magnin. A review of cardiac image registration methods. *IEEE Transactions on Medical Imaging*, 21(9):1011–1021, September 2002. DOI: 10.1109/TMI.2002.804441

[105] M E Phelps, E J Hoffman, and M M Ter-Pogossian. Attenuation coefficients of various body tissues, fluids and lesions at photon energies of 18 to 136 keV. *Radiology*, 117:573–583, December 1975.

[106] J H Ware, F Mosteller, F Delgado, C Donnelly, and J A Ingelfinger. P values. In J C Bailar III and F Mosteller, editors, *Medical Uses of Statistics*, pages 181–200. NEJM Books, Boston, MA, second edition, 1992.

[107] J M Michalski. Neuroblastoma. In C A Perez, L W Brady, E C Halperin, and R K Schmidt-Ullrich, editors, *Principles and Practice of Radiation Oncology*, pages 2247–2260. Lippincott Williams and Wilkins, Philadelphia, PA, 4th edition, 2004.

[108] E E Chaffee and E M Greisheimer. *Basic Physiology and Anatomy*. J. B. Lippincott Company, New York, NY, third edition, 1974.

[109] G J Tortora. *Principles of Human Anatomy*. John Wiley & Sons, Inc., Hoboken, NJ, tenth edition, 2005.

[110] E Keatley, G Mageras, and C Ling. Computer automated diaphragm motion quantification in a fluoroscopic movie. In *Proceedings of the 22nd Annual International Conference of the IEEE Engineering in Medicine and Biology Society*, pages 1749–1751, Chicago, IL, July 2000. DOI: 10.1109/IEMBS.2000.900422

[111] R Beichel, G Gotschuli, E Sorantin, F Leberl, and M Sonka. Diaphragm dome surface segmentation in CT data sets: A 3D active appearance model approach. In *Proceedings of the SPIE International Symposium on Medical Imaging: Image Processing*, volume 4684, pages 475–484, San Diego, CA, May 2002.

[112] X Zhou, H Ninomiya, T Hara, H Fujita, R Yokoyama, H Chen, M Kanematsu, and H Hoshi. Automated identification of diaphragm in non-contrast torso CT images and its application to computer-aided diagnosis systems. In *Proceedings of CARS 20th International Congress and Exhibition: Computer Assisted Radiology and Surgery*, pages 366–367, Osaka, Japan, 2006.

[113] M N Prasad and A Sowmya. Detection of bronchovascular pairs on HRCT lung images through relational learning. In *IEEE International Symposium on Biomedical Imaging: Nano to Macro*, volume 2, pages 1135–1138, Arlington, VA, April 2004. DOI: 10.1109/ISBI.2004.1398743

[114] T Zrimec and S Busayarat. A system for computer aided detection of diseases patterns in high resolution CT images of the lungs. In *20th IEEE International Symposium on Computer-Based Medical Systems*, pages 41–46, Washington, DC, June 2007. DOI: 10.1109/CBMS.2007.13

[115] M S Brown, J G Goldin, M F Mcnitt-Gray, L E Greaser, A Sapra, K-T Li, J W Sayre, K Martin, and D R Aberle. Knowledge-based segmentation of thoracic computed tomography images for assessment of split lung function. *Medical Physics*, 27(3):592–598, March 2000. DOI: 10.1118/1.598898

[116] X Zhou, T Hayashi, T Hara, H Fujita, R Yokoyama, T Kiryu, and H Hoshi. Automatic segmentation and recognition of anatomical lung structures from high-resolution chest CT images. *Computerized Medical Imaging and Graphics*, 30(5):299–313, August 2006. DOI: 10.1016/j.compmedimag.2006.06.002

[117] S Hu, E A Hoffman, and J M Reinhardt. Automatic lung segmentation for accurate quantitation of volumetric x-ray CT images. *IEEE Transactions on Medical Imaging*, 20(6):490–498, 2001. DOI: 10.1109/42.929615

[118] J Langman and M W Woerdeman. *Atlas of Medical Anatomy*. W B Saunders Company, Philadelphia, PA, 1978.

[119] H Gray. *Anatomy of the Human Body*. Philadelphia: Lea & Febiger, New York, NY, 20th edition, May 2000.

[120] R A Zoroofi, Y Sato, T Sasama, T Nishii, N Sugano, K Yonenobu, H Yoshikawa, T Ochi, and S Tamura. Automated segmentation of acetabulum and femoral head from 3-D CT images. *IEEE Transactions on Information Technology in Biomedicine*, 7(4):329–343, December 2003. DOI: 10.1109/TITB.2003.813791

[121] L M Ellingsen and J L Prince. Deformable registration of CT pelvis images using Mjolnir. In *Proceedings of the 7th Nordic Signal Processing Symposium, 2006 (NORSIG 2006)*, pages 46–49, Reykjavik, Iceland, June 2006. DOI: 10.1109/NORSIG.2006.275273

[122] J Pettersson, H Knutsson, and M Borga. Automatic hip bone segmentation using non-rigid registration. In *18th International Conference on Pattern Recognition (ICPR'06)*, volume 3, pages 946–949, Los Alamitos, CA, 2006. IEEE Computer Society. DOI: 10.1109/ICPR.2006.299

[123] C F Westin, S Warfield, A Bhalerao, L Mui, J Richolt, and R Kikinis. Tensor controlled local structure enhancement of CT images for bone segmentation. In *Proceedings of the 1st International Conference on Medical Image Computing and Computer-Assisted Intervention*, pages 1205–1212, London, UK, October 1998. Springer-Verlag. DOI: 10.1007/BFb0056310

[124] A Bousvaros, D R Kirks, and H Grossman. Imaging of neuroblastoma: an overview. *Pediatric Radiology*, 16:89–106, 1986. DOI: 10.1007/BF02386629

[125] H N Caron and A D J Pearson. Neuroblastoma. In P A Voûte, A Barrett, M C G Stevens, and H N Caron, editors, *Cancer in Children: Clinical Management*, pages 337–352. Oxford University Press, Oxford, UK, 2005.

[126] J L Grosfeld. Risk-based management of solid tumors in children. *The American Journal of Surgery*, 180:322–327, November 2000. DOI: 10.1016/S0002-9610(00)00489-X

[127] G M Brodeur, R G Seeger, A Barrett, F Berthold, R P Castleberry, G D'Angio, B DeBernardi, A E Evans, M Favrot, A I Freeman, G Haase, O Hartmann, F A Hayes, L Helson, J Kemshead, F Lampert, J Ninane, H Ohkawa, T Philip, C R Pinkerton, J Pritchard, T Sawada, S Siegel, E I Smith, Y Tsuchida, and P A Voûte. International criteria for diagnosis, staging, and response to treatment in patients with neuroblastoma. *Journal of Clinical Oncology*, 6(12):1874–1881, December 1988.

[128] R H Vu, R M Rangayyan, and G S Boag. Multi-seed segmentation of the primary tumor mass in neuroblastoma using opening-by-reconstruction. In *Proceedings of the IASTED International Conference on Biomedical Engineering*, pages 242–249, Innsbruck, Austria, 2006.

[129] C E Metz. Basic principles of ROC analysis. *Seminars in Nuclear Medicine*, 8(4):283–298, 1978. DOI: 10.1016/S0001-2998(78)80014-2

About the Authors

SHANTANU BANIK

Shantanu Banik is currently a Ph.D. student at the University of Calgary, Calgary, Alberta, Canada. He received his M.Sc. degree in 2008 in Electrical and Computer Engineering from the University of Calgary, Calgary, Alberta, Canada and his B.Sc. degree in 2005 in Electrical and Electronic Engineering from the Bangladesh University of Engineering and Technology (BUET), Dhaka, Bangladesh. He is working on the problem of detection of architectural distortion in prior mammograms to aid the process of early detection of breast cancer. His research interests include medical signal and image processing and analysis, landmarking and segmentation of pediatric computed tomographic (CT) images, and automatic segmentation of the primary tumor mass in children with neuroblastoma.

RANGARAJ M. RANGAYYAN

Rangaraj M. Rangayyan is a Professor with the Department of Electrical and Computer Engineering, and an Adjunct Professor of Surgery and Radiology, at the University of Calgary, Calgary, Alberta, Canada. He received the Bachelor of Engineering degree in Electronics and Communication in 1976 from the University of Mysore at the People's Education Society College of Engineering, Mandya, Karnataka, India, and the Ph.D. degree in Electrical Engineering from the Indian Institute of Science, Bangalore, Karnataka, India, in 1980. His research interests are in the areas of digital signal and image processing, biomedical signal analysis, biomedical image analysis, and computer-aided diagnosis. He has published more than 130 papers in journals and 210 papers in proceedings of conferences. His research productivity was recognized with the 1997 and 2001 Research Excellence Awards of the Department of Electrical and Computer Engineering, the 1997 Research Award of the Faculty of Engineering, and by appointment as a "University Professor" in 2003, at the University of Calgary. He is the author of two books: Biomedical Signal Analysis (IEEE/ Wiley, 2002) and Biomedical Image Analysis (CRC, 2005); he has coauthored and coedited several other books. He was recognized by the IEEE with the award of the Third Millennium Medal in 2000, and was elected as a Fellow of the IEEE in 2001, Fellow of the Engineering Institute of Canada in 2002, Fellow of the American Institute for Medical and Biological Engineering in 2003, Fellow of SPIE: the International Society for Optical Engineering in 2003, Fellow of the Society for Imaging Informatics in Medicine in 2007, and Fellow of the Canadian Medical and Biological Engineering Society in 2007. He has been awarded the Killam Resident Fellowship thrice (1998, 2002, and 2007) in support of his book-writing projects.

GRAHAM BOAG

Graham Boag was born in Montréal, Québec, Canada, on June 6, 1955. He received a Bachelor of Arts (with distinction) in biology at Queen's University in Kingston, Ontario, Canada, in 1978, followed by Doctor of Medicine from Queen's University at Kingston in 1982. His further training included residency training in diagnostic imaging at Queen's University from 1983 until 1987, as well as clinical fellowship training in pediatric radiology at the University of Toronto, Toronto, Ontario, Canada, from 1987 to 1988, and in magnetic resonance imaging at the Children's Hospital of Eastern Ontario in 2003 to 2004.

Graham Boag is a Clinical Associate Professor in the Departments of Radiology and Paediatrics at the University of Calgary, Alberta, Canada. He is a pediatric radiologist and former Department Head in Diagnostic Imaging at the Alberta Children's Hospital in Calgary, Canada. His research interests are in the area of computer-assisted diagnosis and image enhancement in the field of diagnostic medical imaging. He is currently involved in research projects in association with the Department of Electrical and Computer Engineering at the University of Calgary.

Index